Platform and Collective Intelligence

Series Editor
Fabrice Papy

Platform and Collective Intelligence

Digital Ecosystem of Organizations

Antoine Henry

iSTE

WILEY

First published 2021 in Great Britain and the United States by ISTE Ltd and John Wiley & Sons, Inc.

ISTE Ltd
27-37 St George's Road
London SW19 4EU
UK

www.iste.co.uk

John Wiley & Sons, Inc.
111 River Street
Hoboken, NJ 07030
USA

www.wiley.com

Library of Congress Control Number: 2020946598

British Library Cataloguing-in-Publication Data
A CIP record for this book is available from the British Library
ISBN 978-1-78630-492-6

Contents

Introduction

The scientific work and research that concerns organizations has grown considerably in recent years. This is due to the polarization of human activities within structures whose projects and vocations are extremely diverse. It is then as many pluriverses and plurivocal processes that coexist, function and interact locally, globally – and even *glocally* – in the multitude of these existing organizations and social structures whose legal and managerial form often remains conventional (Svensson 2001; Batazzi-Alexis 2002; Eberhard 2013; Koop 2018). Moreover, this polarization in organizations has spread with the unifying framework represented by the "information society" (IS), which has been further aggravated by the intensification of the economic dimension in contemporary societies (Ahrne and Brunsson 2005). This recent context has exacerbated the place of organizations, identified as atomic entities through which, and within which, a wide range of the multiplicity of human activities is expressed. The complexity of these agencies, which are being developed on a planetary scale with the explosion of international economic transactions, explains why the organizations are studied by a wide variety of disciplines and scientific fields that are often complementary: management sciences, economics, information technology, law, commerce, sociology, anthropology, social psychology, ethnology, and information and communication sciences (Geslin 2002; Bernard 2004).

It is clear that today's organizations, which have risen to the new "information society", much to the delight of political and economic forces, bear witness to a new complexity engendered by a heterogeneous set of structures whose boundaries have become more flexible and whose communication processes have become intertwined. These structures increasingly interact through an increase in self-emerging cross-cutting collaborative activities and processes (El Amrani 2008). This complexity surpasses the binary categorization used until recently, distinguishing according to a trivial typology between private and public entities: the former are generally associated with the world of business, trade and industry, and the latter with the institutional and governmental world in its national or international declinations (such as the UN, UNESCO, ISO and WHO) (Bouillon 2005; Bernard 2016), both evolving until now in an implicitly recognized territorialization.

The massive penetration of information and communication technologies (ICTs) in all spheres of human activities and the unprecedented expansion of digitally mediated communications have completely reshuffled the maps and functional boundaries of these basic types of organizations built on a common hierarchical decision-making model. An unprecedented complexity has emerged, which is driven by the immediate concentric and/or eccentric influences propagated by the dissemination and sharing of natively digital information. Concentric influences emanate from outside the organization and intrude, either voluntarily or accidentally, into the internal processes of organizations outside the usual and regulated channels of information dissemination. Eccentric influences, on the other hand, are those that an organization exerts on the outside, voluntarily or accidentally, through the same technological channels for disseminating and sharing information, as well as outside the usual and regulated circuits defined by the organization. This sudden porosity – of an informational and communicational nature – of the clear boundary that recently separated the inside of an organization from the outside, has come about with the massive adoption and exploitation of these computer technologies specific to the Internet, the Web of Documents and the Social Web (Navarro 2001; Gardey 2003; Greenan and Mairesse 2006; Hochereau 2006; Walkowiak 2006).

The technological tools, inspired by the Internet and the early Web (Web 1.0), which became widespread in all organizations, from the smallest to the largest multinationals, brought much more than operational comfort to organizations. Previously, their decision-makers were worried about the inevitable time to upgrade their IT systems, given the impact in terms of cost, reorganization of practices and training of staff, and were not at ease with the procedures for re-computing, upsizing or migrating their information systems (Botta-Genoulaz 2007; Legrenzi 2015; Besson 2016).

The widespread use of Web technologies, supported by the World Wide Web Consortium and its powerful members, has definitively dispelled doubts about the operational efficiency, robustness, scalability and adaptability of these digital technologies (Papy and Sansonetti 2014). Major technological achievements such as YouTube, Amazon, Instagram, Twitter and Facebook, in their respective economic registers, retain their leadership role and, through their technological and economic performance, support the technological architectures on which their success was built. These extraordinary digital devices, via their platform marketplace with a global audience, drive a large part of the world's economic activities and are also the broadcasters/recipients of a considerable proportion of online digital traffic. These spectacular technological and economic successes are now inspiring the most recent platforming projects, to which all private and public organizations are committed, whatever their scope of action. It is still the same digital technologies that are being mobilized and acclaimed, contributing to the increase, if there was still a need, of their trust capital.

Thus, the performative efficiency of Web technologies and their rapid acceptability, both by the technical teams that implement them and by the users who have to integrate them into their work processes, have determined the decision-makers and information technology departments of these organizations to adopt these technologies, which have been instrumental for two decades now in the Web of Documents, the Web of Data, the Semantic Web, the Social Web and the Commercial Web.

The deployment of Web technologies within organizations is now confirmed and applies, through specific applications, to all processes essential to the operation of organizations: administrative, accounting, logistics, commercial and industrial. However, this deployment is not insignificant (Hénnocque 2002; Rifkin 2002; Balpe *et al.* 2003; Berry 2008; Doueihi 2008). Web technologies, heir to the conceptual and technical principles of the Internet, convey all the meshing and networking mechanisms for which the

TCP-IP communication protocol and its associated historical services (such as RCP, RCMD, FTP, SMTP, NNTP, WAIS, GOPHER and HTTP) have been developed. The social, participative and community evolution that was imposed with the first peer-to-peer distributed applications has continued to grow with digital social networks where hundreds of millions of users around the world interact (De Gail 2013). The reticular logics of communication and the diffusion of information have propagated in organizations at the same speed as the deployment of software environments that have switched infocenters and groupware to the intranet, cloud computing and Software as a Service (SaaS) to reach its most sophisticated expression: digital platforming.

Computer applications using Web technologies are now homogenizing infocommunication practices regardless of the size and nature of the organization: multinational, government, institution, NGO, association, family business or craftsman. The evolution of digital practices and uses that have spread massively with mobile telephony has finished consuming the informational and communicational porosity of organizations. The perpetual rearrangement of communication, sharing and dissemination of information within organizations, towards the outside world and according to hybrid internal/external formats, are reconstituting virtual communities of interests or practices that bring together users who are freeing themselves from traditional and established hierarchical logics (Henry 2019a).

Chapter 1 proposes to return to the inclusion of organizations within this conception of the "information society" by contextualizing notions related to the digital world, organizations and relations with users. We shall return more particularly to the question of the inherent temporality of digital technology, and then focus on the mediation provided by technical devices, before illustrating the transformations brought about by digital technology.

Chapter 2 explores the issue of the platform organization in order to position different notions (such as the network organization) to deal with the platformization of organizations and the particular case of the State as a platform.

Chapter 3 is devoted more specifically to the formation of the sociotechnical instruction of collective intelligence. To this end, we will study the way in which the social is instrumented by the digital, as well as the vision carried by both the sociotechnical and the operational model of the transformation of the organization linked to collective intelligence.

Finally, we will come to the notion of "organizational experience" by focusing, in Chapter 4, on the capacity of actors to transform their organization by mobilizing the concepts explored in the previous chapters.

1

Organizations and Digital Technology

Within contemporary society, information and communication technologies (ICTs) are the bearers of the "imaginary of a technical action initiating new social projects" (Papy 2009, p. 15).

In recent years, public and private organizations[1] have been receptive to this discourse and have equipped themselves with digital technical devices that respond to the new forms of modernity brought about by ICT and the "information society" (IS).

In order to achieve the ambition of this book, it seems useful to present the current context of Western organizations and more particularly that inherent to the information society.

---

1 Organization is broadly understood as "a set of social, legal, technical, cultural, economic and ethical arrangements designed to enable socially organized actions to unfold in time and space to achieve results" (Lorino 2005, p. 55). Organization is then a process that is built through the actions of its members or communities of practice and that Bouillon calls "organization-process" (Bouillon 2015). From this point on, we are in an information approach of the organization by accepting the fact that "an organizational process thus implies a flow of information that makes it possible to articulate material flows and activity flows within the organization" and that "the logic of information processing influences, indeed largely determines, that of activities, since the physical, material and social organization must become as 'fluid' and 'agile' as the flow of information that accompanies the activity flows" (Bouillon 2015, p. 93). The organization here allows for the contextualization of action, the creation of meaning and values for its members.

1.1. Organizations of the "information society"

Having emerged in the 1990s, the concept of the "information society" reflects a conception that combines political[2] discourse and technical[3] infrastructure. The whole is marked by neo-liberalism in a context of accelerated globalization and strong economic competition. Thus, in three decades, the digitization of society has materialized, mainly in the Northern countries[4]. This social project, carried by industrial countries and private[5], public[6] or associative actors, is then "a source of new forms of organization and emancipation, it is also a source of new modalities of social control and of enslavement" (Mattelart 2018, p. 4). In fact, technology, instantiated by digital technologies, no longer expresses itself only in control or structuring, but accompanies the social construction of society by giving it a technical foundation. Castells (1998) presented this vision in his reference work *La société en réseau*, as did Rheingold (2005, p. 16), for whom "the greatest successes in the information and communication technology (ICT) industry will come not from hardware or software, but from new social practices". Thus, ICTs can extend to the entire social structure (Castells 1998, p. 525).

2 Like the "information superhighways" mentioned by US Vice President Al Gore or the plans put in place in France, such as the "France numérique 2012–2020" plan or the portal for the digital transformation of businesses, the *portail de la transformation numérique des entreprises* (https://www.francenum.gouv.fr/) (all the links in the book were available on 01/02/2020).

3 For example, fiber or mobile phone standards. These technologies, combined with the competitiveness of the country and territories, are being generalized in France. For example, the "*Très Haut Débit*" (Very High Speed Broadband) mission (https://www.francethd.fr/) should cover the entire country by 2022. The initial investment was 20 billion euros over 10 years, but the *Cour des comptes* estimates that the project should exceed the 35 billion euro budget and last until 2030.

4 In 2018, in the United States, among the 10 most highly valued companies on the stock exchange, eight of them are from the digital sector: http://www.leparisien.fr/economie/business/apple-amazon-facebook-les-10-entreprises-les-mieux-capitalisees-en-bourse-03-08-2018-7842417.php.

5 The major companies of the technological world and in particular those grouped under the acronyms GAFAM for Google, Apple, Facebook, Amazon and Microsoft (in the United States) or BATX for Baidu, Alibaba, Tencent and Xiaomi (in China) are particularly concerned about this project of a digital society.

6 For example, the FrenchTech label (https://www.lafrenchtech.com/fr/) awarded by the mission of the same name and which was created in 2014 by Axelle Lemaire, Secretary of State for Digital technology, to promote French digital start-ups.

The inclusion of the organization in the "information society", where information is an essential economic asset to be competitive (Miège 2002; Mevel and Abgrall 2009; Ng 2013), is a trend that can be illustrated by the most active sectors in terms of the number of patent applications filed (Figure 1.1).

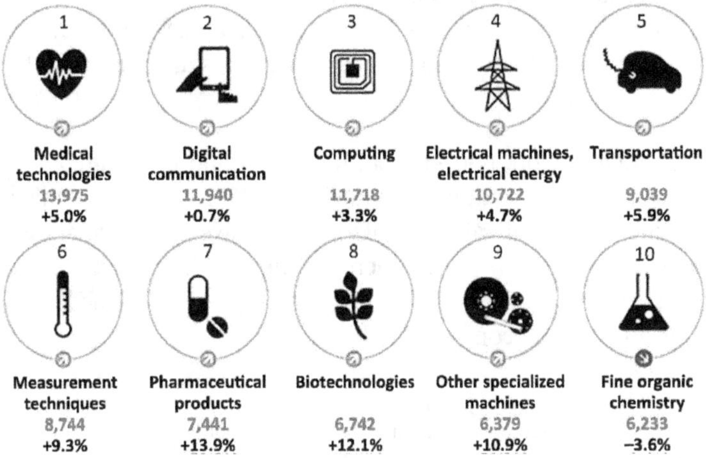

Figure 1.1. *Most active technology sectors in 2018 in terms of number of patent applications (source: European Patent Office, 2019)*

COMMENT ON FIGURE 1.1.– *While some sectors are directly related to information and its processing (notably the digital communications and computer sector), the transport or health sectors are also major players today in relation to the issues of connected health or autonomous vehicles. Information, in its digital form, then becomes central to the development of these economic sectors.*

This informational orientation makes it possible to reposition the organization around the "information culture[7]" from the moment that it is accepted that there is "no information society without information culture" (Baltz 1998, p. 77).

7 The notions of "information culture" and "informational culture" are derived from the notion of "information literacy". For more details on the concept, see Serres (2009a, 2009b).

This point of view is shared by Guyot, for whom information "is part of our daily environment – we use it, we seek it, we produce it at all times – which classifies it as one of those gestures that are so inconsistent that it seems inappropriate to wonder about them" (Guyot 2006, p. 12). Today, information is taken into account in the production of goods or services (Guyot 2006, p. 12), and also as a competitive advantage in the context of the implementation of strategic intelligence (Castagnos and Lesca 2004; Moinet 2009).

It is this phenomenon, both present in society and in organizations, which is called "informationalization" by Miège (2007). He describes this notion as "the increasing and accelerated circulation of flows of information, whether published or not, in the private sphere, in the work environment or in the public space" (Miège 2007, p. 66). The latest data published by INSEE[8], illustrated in Figure 1.2, confirm this trend.

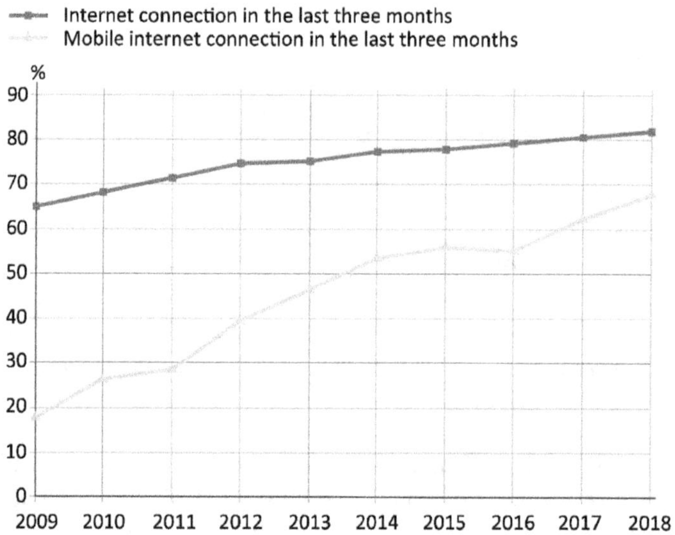

Figure 1.2. *Percentage of people having connected to the Internet in the last three months according to the INSEE survey on ICT use by households between 2009 and 2018 (source: INSEE)*

8 See the results of the survey "L'usage des technologies de l'information de la communication par les ménages entre 2009 et 2018" published in March 2019: https://www.insee.fr/fr/statistiques/3743352?sommaire=3743360#consulter-sommaire.

COMMENT ON FIGURE 1.2.– *In light of these results, ICT use continues to grow in France in all segments of the population studied by INSEE. However, disparities in use remain, particularly between age groups and according to urban/rural location. Internet access has grown very strongly with the introduction of the so-called "smartphones" and with the introduction of 3G, then 4G and soon 5G, which promote mobile Internet access.*

The devices associated with this "information society" lead to an increasing instrumental mediation, both in private life and in the professional context, for access to information, one of the fundamental dimensions of which is temporality.

1.2. Temporality of devices and organizations

In a collective, temporality is a source of intelligibility, i.e. it participates in the understanding of human phenomena within organizations (Dubar and Rolle 2008). Time even conceals a "capacity to define, to a certain extent, the identity of the actors, [and] also has the capacity to define the 'reality' of their socio-cultural environment" (Semprini 2003, p. 53). For Semprini, this is an essential notion for accurately analyzing an actor or a situation to the point where, according to him, temporal logic unfolds as the current paradigm of the "flow society".

1.2.1. *Internal temporality of the devices and temporality of the organization*

In the devices, several temporalities cohabit and sometimes oppose each other. The technical systems are part of a logic of flows and processes in which ICTs promote the "liquidity" of modern society, i.e. a society "where the conditions in which its members act change in less time than it takes for modes of action to become fixed in habits and routines" (Bauman 2006, p. 7). Relationships, routines and knowledge are constantly evolving to meet the imperatives of this "liquid" society.

Alongside these considerations directly related to ICTs, Aubert (2003) and Finchelstein (2011) argue that society and organizations are in a state of urgency leading to a permanent tension of adaptation to societal expectations. This consideration leads to the acceptance of acceleration as "the major experience of modernity" (Rosa 2013, p. 1). Rosa confirms this experience by isolating three axes of acceleration: technological acceleration, the acceleration of social change and the acceleration of the pace of life. Here, he explicitly states the influence of technology on the question of time. Thus, the basis of the relationship to current time "lies in the alliance that has been forged between the logic of immediate profit, that of the financial markets, which reign supreme over the economy, and the immediacy of the new means of communication" (Aubert 2003, p. 24). The situation that is taking shape leads to the domination of "real time"; a tyranny of the moment, for Carayol and Bouldoires (2011), where everything is just-in-time and in a logic of urgency – even hyper-reactivity – permitted by ICTs.

This conception of organization, marked by ICTs, exacerbates the problematic of temporality in organization (Thoemmes 2008; Domenget *et al.* 2017; Lépine *et al.* 2017). It is a reflection that is now significant in relation to the acceleration imposed by ICTs[9], one characteristic of which is to "accelerate exchanges" (Kiyindou 2016, p. 13).

Technological tools are reshaping the perception of time, because "the media are mental operators, acting in the construction of our representations and particularly in the representation of time; in this process, the media play an important role, directly proportional to the size of society and the complexity of social organization" (Martino 2011, p. 41). They participate in the construction of the temporality of the action and its materialization through the mediation imposed by the device.

1.2.2. *Permanent prototype deployment*

In keeping with the immediacy of the world, organizations in general and businesses in particular are currently under pressure to adapt almost instantaneously to meet user expectations. However, each activity carried out within the organization has its own temporality.

9 As tools contributing to our representation and the division of time.

Considering that the development of the device is linked to its use[10], it finds itself in a situation that can be described as "unfinished by design" (Garud *et al.* 2008), i.e. "the instability, lability and unfinished nature of devices today constitute a sustainable state" (Domenget 2013, p. 43). This permanent non-completion questions the organization's capacity to adapt in real time to its external environment (responses to market or legal stimuli), as well as internally to the practices and uses of its members.

The system remains incomplete from the point of view of uses that are still evolving and unstable; this means that a great deal of attention must be paid to real, current uses and no longer potential or imagined uses. The next step is to "bring back" these uses, to consider them as a major source of inspiration for design policies (Gaglio 2010). These uses, which are "implicit demands", are encountered by developers by working iteratively (trial and error) and testing their "imagination of possible uses" (Papy 2016, p. 32).

Based on this observation, in the case of the development of the schemes, "it is not possible to consider the frameworks as pre-existing. They are to be considered at the same time as the action itself" (Aggeri 2014, p. 48). This consideration is fully in line with the logic of the term "beta"[11], which

10 Thus, we will use the term "usage" if it concerns the individual framework and that of practice for activities concerning a community of practice or the entire organization, considering that "usages give meaning to technical objects by inserting them into pre-existing practices" (Le Marec 1997, p. 540). Concerning the notion of use, as Chaudiron and Ihadjadene (2010, p. 3) point out, "whereas the term 'use' refers to the way in which a particular user, in his or her own singularity, discovers, apprehends and manipulates a device (such as a search engine, for example), the notion of use has a broader meaning". On practices, and more particularly digital practices, Agostinelli (2009) states that their analysis aims to "identify the fundamental characteristics (social functions, corresponding situations and problems, fundamental concepts, specific strategies, etc.) and thus define invariants of a digital culture, research on practices is constituted in transversal knowledge based on field work and recognized by the interdisciplinarity stemming from the various fields of publication. They constitute themselves as actors of a paradigm built on lived experience and their involvement in the information society" (Agostinelli 2009, p. 357).

11 This is a fundamental characteristic of Web technologies that is being transposed in organizations today. Indeed, in a computer development cycle, the software goes from the alpha version (close to the prototype) to the beta version (functional but not finished software) then to the release candidate version (software almost finished) and finally to the stable version.

symbolizes a state of completion sufficient to make it, in the case of software, testable as is, while accepting its incompleteness (e.g. errors in the computer code or an unfinished human-machine interface). It is then conceivable to develop the system according to the action situations in which it is used, or even to rethink it in its entirety in order to end up with a finalized product adapted to the needs.

This leads to a situation of "flexible interpretive technologies" (Feeberg 2014, p. 43), which states that the design of a device cannot be completed until the uses are well defined. This situation then makes the analysis of uses more complex, as they are constantly in the process of stabilization and instability.

By their "permanent beta" or "perpetual beta" dimension (Gallezot and Le Deuff 2009; Passant 2009; Papy and Sansonetti 2014), ICTs convey their own temporality. Moreover, for an organization, the coherence of its IT assets, linking both old information systems and new technologies, represents a complex situation with several categories of costs: feasibility costs, acceptability costs and costs of compatibility between applications (Benghozi 2000). This complexity contributes to discrepancies between, on the one hand, organizations with a long history of sometimes old software and, on the other hand, the expectations of their employees who implement the most recent ICT in their private sphere (Boboc and Metzger 2009).

1.3. ICT-enabled mediation

The digital, perceived as an antechamber of the real (Bordier *et al.* 2012)[12], mediates with reality in the sense that "an individual never interacts directly with the environment. The relationship between the individual and the objects of the environment is mediatized through means, tools and cultural signs" (Lewandowski and Bourguin 2009, p. 159). In fact, the action mediated, here by digital technology, does not allow the actor to directly access the real, which is then part of the triptych reproduced in Figure 1.3.

---

12 Bordier *et al.* (2012) illustrate this aspect through the use of 3D technologies or virtual/increased reality that model reality.

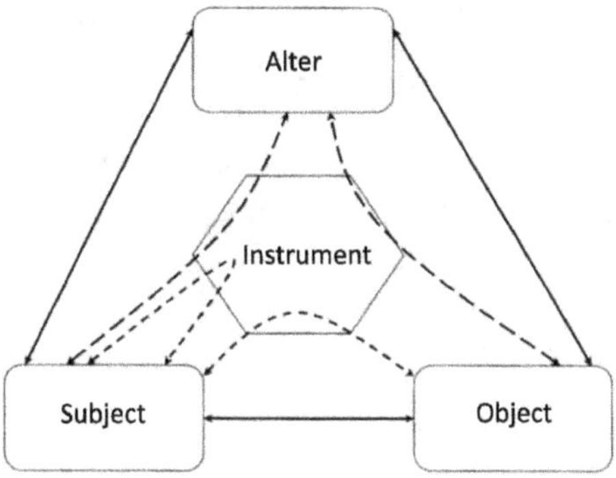

Figure 1.3. *Direct and media reports supported by the instrument (according to Rabardel 1995a, p. 53)*

COMMENT ON FIGURE 1.3.– *This is a schematic representation that illustrates how the subject passes through the instrument to reach reality or his or her own perception of the instrument. The digital traces (Vayre 2014; Beckouche 2017) are thus an example of what an instrument can generate for a subject; the subject generating his own digital trace through the digital instrument or apprehending an object through it. What makes it possible to distinguish one activity from another is "the difference between their objects, because it is the object of the activity that gives it its orientation" (Rabardel 2005, p. 252). Digital technology, by virtue of its role as a mediator, helps to produce this mediatized activity, here in the framework of a private organization, and offers the possibility of using theories of action to analyze them.*

Digital technology is therefore a tool that materializes concretely in the form of technical devices that translate and even interpret reality[13]. Indeed, in the wake of mathematicians such as Leibniz, the digital constructs a representation of the world that is mathematicized (Mattelart 2018). Whether it is the individual, or his/her actions, social relations or knowledge, everything is calculated and represented from this calculation (Carmes and Noyer 2014), so that software predominates in our society by extracting value and contributing to the representation of the world (Galloway 2012).

13 Of which the example of augmented reality is here a possible illustration. Indeed, the technical device adds, in this case, an informational overlay to a real situation.

However, this means of action, whether tangible or intangible, is a constitutive effect of the subject's action. Rabardel (2005) completes his remarks by specifying that instruments[14], like the artifacts of the "information society", are not neutral or value-free. The said instruments then participate in the construction of reality, with users implementing "devices that influence activities and individual representations" (Bouillon *et al.* 2008, p. 5). Through these infocommunication phenomena, the devices are associated with the construction and individual and collective implementation of meaning or values.

The use of technical devices, such as office software or digital social networks, has become a daily prerequisite for carrying out individual or collective tasks in both private and public contexts. We therefore find the triplet "individual–technology–organization" (Bernard 2000, p. 36) where the technology carries out the function of mediation between the individual and the organization, so as to consider the use of the concept of mediation by putting it into a practical and organizational perspective.

The origin of this transformation dates back to the end of the 20th Century. Indeed, computer technology was introduced, in particular, by computerized management tools, as illustrated by Pavé (1989). The purpose of these tools is to facilitate the rationalization of the organization. Indeed, with computerization, it is not only the information circuits but also the activities of men and women that are organized. In this information technology, which we can describe as business informatics (Hochereau 2006; Delcambre 2017), we find tools such as Enterprise Resource Planning (ERP), similar to "organizational technologies" (Besson 2016, p. 18) leading to organizational change.

This phenomenon is clearly observed among industrial players. In the early 2000s, three researchers – Pierre-Jean Benghozi, Patrice Flichy and Alain d'Iribarne (Benghozi 2000) – illustrated the breakthrough of ICTs in organizations looking for performance gains, a trend that is not being

14 Rabardel accepts the instrument as "a mixed entity that includes, on the one hand, the material or symbolic artifact and, on the other hand, the usage patterns, the representations that are part of the user's skills and are necessary for the use of the artifact. It is this mixed entity, which derives from both subject and object, that constitutes the true instrument for the user" (Rabardel 1995b, p. 64).

questioned today despite examples of misuse of devices. Thus, ERP controls the actor by cutting and organizing his/her time to achieve objectives that are "quantified, measured and recorded" (Dujarier 2017, p. 18); it becomes a "forcing system" where the system no longer fits into the structure of the organization but transforms it to adapt to its functioning (Meyssonnier and Pourtier 2004).

This reflection on organization leads us to return to the place of technical devices in its functioning (both in meeting the primary needs of the organization such as communication between its members and in a logic of optimization). Through the political choices that are made, the system favors a specific organization and operating mode linked to it. Indeed, by mediating the action, the technical device influences knowledge, know-how and even its use (need for specific knowledge, structuring of human relations). As a result, the devices used by human beings to mediate their relations with the world become predominant to the point where technology is positioned as the "single mediator" according to Akrich (1993).

1.3.1. *From IT to ICT*

A certain amount of vagueness still exists today between the multiple terminologies that can be used, such as computer science and ICT, or other designations such as information technology (IT).

Simonnot (2009) identifies the development of a computer culture in the 1980s with the appearance of personal computers. This computer culture gradually gave way to digital culture and ICT. In fact, at the end of the 1980s, information technologies and telematics were brought together to move towards ICTs, a situation that then initiated the transition from computer science to digital technology.

With this change, the digital culture, including its functional components that are ICTs, is resulting in a technicization of uses. Web 2.0[15], conceptualized by Tim O'Reilly (2005), is an example of this (Figure 1.4).

---

15 In his online article, O'Reilly details his vision and illustrates the emergence of what he calls "Web 2.0" with examples: https://www.oreilly.com/pub/a/web2/archive/what-is-web-20.html. He also discusses the positioning of the Web as a platform.

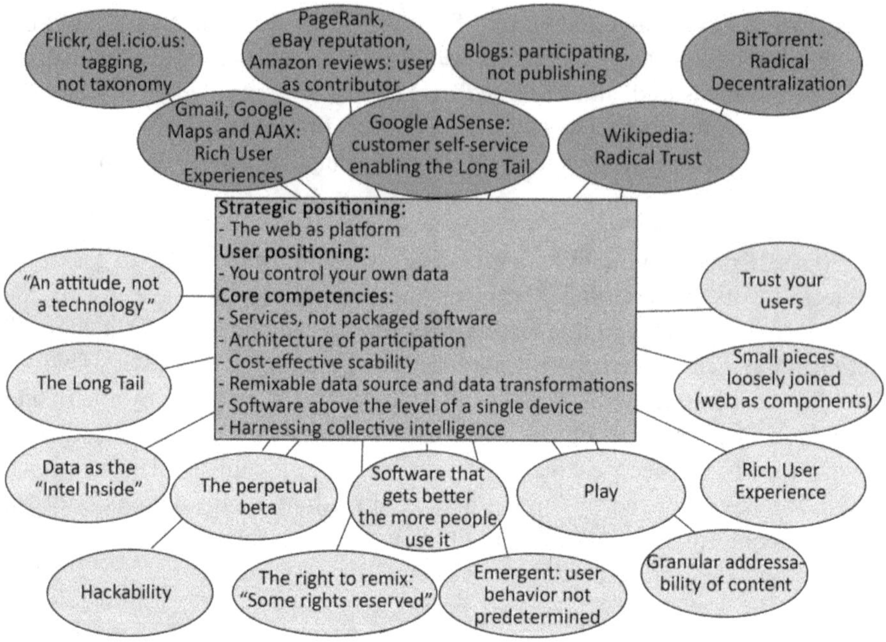

Figure 1.4. *Web 2.0 Meme Map (source: O'Reilly (2005))*[16]

COMMENT ON FIGURE 1.4.– *O'Reilly proposes the* Web 2.0 Meme Map, *based on a joint workshop brainstorming session, to illustrate the changes he identifies with the emergence of Web 2.0. He highlights what he presents as collective intelligence (which is notably expressed in the form of "wisdom of the crowd", which will be discussed in section 3.2) and the capacity of these new tools to mobilize the work of Internet users, also currently known as* digital labor (Cardon and Casilli 2015; Casilli 2015) *(Google, for example, relies on Internet users to solve Captchas and Wikipedia succeeds in mobilizing Internet users to write an encyclopedia).*

Through this "social face of the technological, industrial and political Web, [which] conveys the participatory and collaborative model of sharing and exchange of data" (Papy 2009, p. 20), a "virtual socialization space" (Verlaet 2015) is then established, which takes multiple forms such as digital social

16 https://www.oreilly.com/pub/a/web2/archive/what-is-web-20.html?page=1#mememap.

networks[17] or online games[18]. The different action situations are then mediated by technology.

1.3.2. *Action mediated by technical devices*

The deployment of information technology in social practices and professional activities through ICTs leads to questions about the activity of users, their interactions and the actions they carry out (Flichy 2003).

Mediation through ICTs is a subject abundantly addressed by researchers in the human and social sciences, especially its impacts on the functioning and optimization of an organization in perpetual transformation (Agostinelli 2003, 2009; Le Douarin 2007a; Robert 2009; Patrascu 2010; Tchuente 2010; Pieters 2013). This is concretely reflected in the practices in place, and therefore in the actions of the members of the organization, and this is in line with the various existing technical or social determinisms.

Guyot (2000) identifies three possible forms of mediation:

– *interpersonal*: mediation is the act of another person intervening to carry out an action (without technical mediation therefore). The author highlights the exchange leading to the production of an intercomprehension;

– *supported by a tool*: the interpersonal relationship is then mediated by a technical tool. The use of a tool adds an additional technical activity to the initial action. It is essential that the use of the device is controlled so as not to degrade the quality of the interpersonal relationship;

– *entirely technical*: mediation takes place throughout the action carried out and may even be dependent on technical mediation. This approach requires a significant amount of work, both upstream, in order to design the system

17 Digital social networks use the possibilities offered by Web 2.0 to offer online spaces to Internet users. After pioneers such as MySpace, the major digital social networks today are sites such as Facebook, Weibo, WeChat, Twitter and LinkedIn for professional use.

18 The symptomatic example is the so-called massively multiplayer online role-playing games (MMORPGs) or online games like *Second Life*. These games are social spaces in which Internet users – regardless of their nationality or geographical location – are able to exchange, organize themselves and thus experience digital sociability.

with regard to its target audiences, and downstream, in terms of the skills available to them[19].

The choice of one of these three modes then reveals a political will[20] as well as a representation of technical culture, which is expressed in terms of the relationship maintained with the technical object and its mode of appropriation (Guyot 2000). The act of mediation that is operated by ICTs then makes it possible to question the "link between the actor and the collective, because any mediating action makes it possible to associate and think about the individual in his/her relationship with others and with groups" (Liquète 2012, p. 122).

To illustrate this dimension, a parallel can be drawn with the current trend of focusing on uses to build an information system. By focusing on use, this approach focuses attention on the users of the systems, whose operation must be taken into account by the designers so as to generate as little friction as possible and involve them in the design of their future tool[21]. This approach tries to make a synthesis between the needs of the individual (who is more or less competent to define his/her own uses) and the collective (for whom it is less obvious to be able to decide individually with regard to the uses of others). This trend is currently illustrated by the emergence of the methods known as "agile" (Vallon *et al.* 2017) whose foundations are based in particular on collaboration between future users and the technical team, as well as on the ability of developers to adapt to the changes desired by these users[22].

In each case of production of a device, a specific distribution of competences and a distribution of authority operates. This situation is

---

19 Full dematerialization of public services in France should be achieved by 2021. However, the rights defender warns that "dematerialization offers new means of access to public services and simplifies access to information and administrative documents for a majority of users, but entails a risk of reduced access to rights and exclusion for many of them". The excluded are those who do not have a computer, an Internet connection, specific knowledge related to the use of these technologies, etc. The Advocate's report is available at: https://www.defenseurdesdroits.fr/fr/rapports/2019/01/dematerialisation-et-inegalites-dacces-aux-public services.

20 Since March 1, 2016, Pôle emploi has required the unemployed to register and make their applications via the website rather than at the counters. This decision, which should free up the time of counsellors to accompany job seekers, presupposes that users are equipped with the necessary equipment and Internet connection to access the Pôle emploi website.

21 This is the logic behind approaches such as user experience (UX) or design thinking.

22 The entire Agile Manifesto and its principles can be accessed at http://agilemanifesto.org/.

addressed by Aggeri (2014) who divides the design of a device into several phases:

– *problematization phase*: it consists of producing meaning from the situation to which the device is supposed to respond. This problematization provides information on knowledge needs, on the identification of the problematic situation and on its context;

– *phase of stating a strategic project and proposing a scheme of the device*: by working on the problematization, it is a time of ideation and production of a draft representation of the device;

– *phase of creating spaces for collective action and mobilizing partners*: by mapping the spaces, the forces at play, and the links between the actors, it is possible to target more precisely the needs to be met by the system;

– *instrumentation and scenography phase*: the tool produced must fit into the organization and its context to ensure that the devices form a coherent and non-redundant whole in order to produce the desired effect;

– *scriptwriting and communication phase*: building the collective imagination around the device and the accompanying story.

All of these phases take place more or less simultaneously and accompany the concrete production of the system. Mediation by the device thus depends on the symbolic system of representations that it conveys because "mediatization is to transpose into familiar forms, already experimented, and to solicit associated mental activities" (Guyot 2000, p. 26).

Based on this observation, Folcher (2015) develops two concepts:

– on the one hand, the "designers in use" which are, for a given community, the repositories of the system of representations they share and the experiences of the designers so as to make it a real instrument according to Rabardel's acceptance (1995a, 1995b, 2005);

– on the other hand, the "designers for use" fed by the experts' vision. Located at the organization's head office, far from action situations, these "planners", as presented by Dujarier (2017), transpose financial, organizational or functional objectives, at the risk of designing a system that is not adapted to operational needs.

According to Dujarier, this state of affairs generates four tensions between the vision carried by the device and reality:

1) the discrepancy between the initially foreseen situation and the reality leads to inappropriate use of the device;

2) if the tool focuses on the accomplishment of a task, it is part of a more global context that must be taken into account (preparation time, travel, etc.);

3) for each task, an end is programmed independently of the evolutions encountered during its realization;

4) the user has to achieve a numerical target that is sometimes uncorrelated to reality.

This rationalization and the imperative of the figure are favored by the technical mediation carried out by digital technology. Indeed, the technical system allows for increasing calculability, which leads to a phenomenon of quantophrenia.

1.3.3. *ICT and quantophrenia*

The mediation operated by the technologies of information repose mainly on a mediation nourished by applied mathematics and calculation, a situation inscribed in the "genetic code of the project of society inspired by the mysticism of number" (Mattelart 2018, p. 5).

Indeed, this logic of calculability has become the foundation of this information-based society project. Calculation is a foundation of the "true discourse" and is presented as the outcome of the "quest for the perfection of human societies" (Mattelart 2018, p. 5). Computer science, based on a binary logic, is then a concrete manifestation of this.

Digital actors tend to metaphorize the world, human relationships[23] and behaviors[24]. This is what Doueihi stressed, "digital technology, itself in search of the absolute and total in its effects on people, their societies, their behavior and their values, brings together the internal components of the

23 Thus, Facebook pre-empts the digital "I" and participates in the construction of *extimacy* (a neologism meaning the staging of the intimacy of Internet users outside) to take up the notion developed by Tisseron (2001).

24 The current desire to use predictive algorithms in order to anticipate the behaviors of individuals, whether in their consumption or in their actions, is based on this claim to succeed in modeling the individual through the data he/she generates on the Internet or through his/her behaviors which are digitized to be processed by the said algorithms.

technology in its current state, represented by the platforms as well as by the imaginary of the technology, while allowing flexible uses often shaped by local specificities and which are of a cultural nature" (Doueihi 2011, p. 37).

Finally, the phenomenon currently observable is that if we fail to understand, an effort is made to measure. This state of affairs is illustrated by Muller (2018) in his book *The Tyranny of Metrics* by the shortcomings inherent in this obsession to quantify, measure and compare[25].

Wieviorka underlines in the preface of *Penser global: l'humain et son univers* that "we live in a universe threatened by the sick obsession with numbers and quantitative data, quantophrenia, as Pitrim Sorokin used to say: this should not exempt us from thinking and reflect" (Morin 2015b, p. 77). Thus, quantification and compartmentalization generate fragmented and local knowledge that does not allow a global understanding.

A concrete illustration of this *de facto* situation within organizations – with the dynamic provided by the "ITC-quantophrenia" couple (Pontille and Torny 2013; Ollion and Boelaert 2015) – is the introduction of quantitative social monitoring for the systematic evaluation of employee performance. From a digital labor perspective, individuals are "transformable in resources, in human groups to which they can subcontract tasks" (Casilli 2015, p. 13) and produce, via digital technology, "forms of activities that are akin to work because they produce value, are subject to some form of contractual framework and are subject to metrics of performance" (Casilli 2015, p. 15).

This logic is accompanied by the implementation of a permanent control in order to carry out a systematic evaluation of individuals and employees (Dujarier 2017).

Backed by capitalism, quantophrenia leads to the effective emergence of surveillance capitalism[26] (Zuboff 2019) and thus of "governing by the trace, governing by numbers, algorithmic governmentality" that Mattelart (2018,

25 The author provides in-depth and documented details of situations related to the misuse of measures, i.e. situations leading to abuses and strategies to divert the initial willingness to use different measures (as in the case of the rankings of American universities or hospitals by the *U.S. News & World Report*).

26 The author proposes the following definition of surveillance capitalism: "a form of capitalism that translates human experience into behavioral data in order to produce predictions that are then sold on the market for future behavior" (Zuboff 2019).

p. 96) evokes, amplifying the remarks of Vayre (2015) on "governing by data". Symptomatic of this thought is the assimilation of data to 20th-Century oil[27]. An economic and political conception, data becomes a raw material essential to the good functioning of the so-called "modern" economy. The transition to a data society is linked to a desire to make society predictable[28]. In fact, the society becomes "non-worrying, as well as not informational" (Escarpit 1976, p. 166), because the information is surprising (Gleick 2015, p. 203). The raw data is considered objective, in the sense that it is factual, which can lead to serious abuses in its use. Thus, O'Neil (2018) returns in her book on cases prejudicial to American citizens because of the very absence of a feedback loop to correct the raw data and their errors.

This technical mediation and its consequences take a prominent place in our connection society. This raises the question of the transformation of organizations in order to respond concretely to technological, social and societal changes.

1.4. The transformation of organizations

In order to concretely illustrate the mediation of digital technology, we will deal with the question of the transformation of organizations in relation to digital technology from an organizational point of view first and then from a social point of view, after having introduced elements of understanding of these transformations.

Many authors offer illustrations of technological immanence. Among these, we can note the following works:

– *In the Age of the Smart Machine* (Zuboff 1989): Zuboff illustrates how the factory became "electronic text" and the industry's transition to digital;

27 Nellie Kroes, former Vice-President of the European Commission, frequently used this analogy. Similarly, in this statement on open government data, the European Union refers to this data as "gold": https://ec.europa.eu/commission/presscorner/detail/en/IP_11_1524. In both cases the same extractivist logic applies. This parallel can also be found in the growing contribution of digital technology to climate change in connection with the ever-increasing demand for energy as well as minerals (Pitron 2018).

28 In particular, with the use of predictive algorithms that rely on data to classify, categorize, compute and organize inputs with a view to providing the statistically most likely answer to the initial problem.

– *The Third Industrial Revolution* (Rifkin 2012) (or in his previous essay (Rifkin 2000)): the prospectivist Jeremy Rifkin proposes to qualify as an industrial revolution the use of information and communication technologies in relation to their influence on our daily life. His thinking is also based on the transition from oil (which was partly responsible for economic development in the 20th Century) to data;

– *La société de flux* (Semprini 2003): for Semprini, a new temporal logic has been put in place with the following characteristics: (1) variety and multiplication, time is no longer a monolith, but is segmented and fragmented to adapt to the individual rhythms; (2) time is now indeterminate and open in opposition to the predictable dimension of past temporalities (like the hours dedicated to meals or work). This non-binding temporal logic is adapted to flexibility as opposed to repetitiveness and regularity; (3) the tempo logic in today's society is subjective. Based on these considerations, he comes to develop the notion of flow and flow society as a current paradigm.

Dudezert (2018), in her book on the transformation of organizations in relation to digital technology, presents Figure 1.5, which summarizes the reasons that lead organizations to transform themselves.

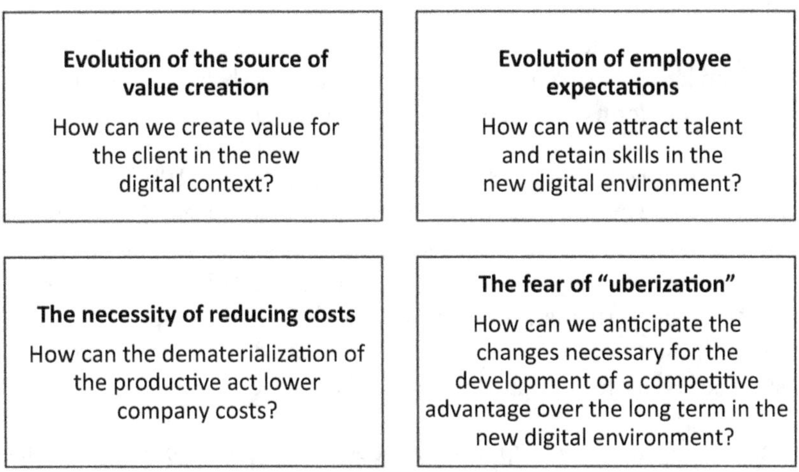

Figure 1.5. *Reasons for organizational transformation (according to Dudezert 2018, p. 59)*

COMMENT ON FIGURE 1.5.– *The author details four reasons why organizations are undergoing digital transformation. (1) The first reason is related to the source of value creation; it refers directly to the assimilation of data to the "oil of the 21st Century". (2) The second reason is inherent to the expectations of employees; it is detailed in section 1.4.2. It illustrates the pressure of those whom Carmes (2010) calls "Intranet–Internet users" and whose increasing personal use is accompanied by demands on the organization and its digital environment. (3) The third reason is in line with the logic of the supposed gain in productivity presented in section 1.4 and the reduction in costs by replacing human operators with machines. (4) The last reason is based on the neologism "uberization". Created in 2014, this term symbolizes the intrusion of a digital player into a regulated sector (such as taxi drivers) to compete with it. Many organizations fear being overtaken in their core business by digital players and are investing heavily in the sector to anticipate this situation.*

Proponents of digital technology highlight the new information processing and analysis capabilities offered to organizations in all areas, cost reductions, opportunities to create a dialogue with clients, as well as to make organizations more efficient by structuring and equipping them. Organizations would then be "aware" of the existence of these opportunities, which is why they would be part of a digital transformation process.

This positive vision is not shared by all researchers, whose critical analyses invite us to put into perspective the benefits of the digital transformation in terms of real and supposed uses.

Authors such as Rallet (1986), Brousseau and Rallet (1997), Greenan and L'Horty (2000), Dumas *et al.* (2017) and Mohamad *et al.* (2017) highlight that the use of technologies in organizations leads to a standardization of tasks, or even a taylorization of work. Rasse (2015) notes, in fact, that taylorization (in particular, the division of labor) is amplified by digital technology[29]. At the same time, there is a strengthening of the control and regulation necessary to continue to develop the division of tasks.

At the same time, the supposed productivity[30] gain linked to the use of digital technology is being called into question (Guyot 2006, 2013). The digital world is not free of the difficulties it creates in terms of uses, in

29 This taylorization echoes digital labor, mentioned earlier, which pushes this logic to its paroxysm by encouraging the division into micro-tasks of a few seconds.

30 Also called the "competitiveness paradox" by Thouin *et al.* (2008) or Yuhn and Park (2010).

particular: "non-use of ICTs [...], digital gaps and fractures [...], attacks on the privacy of individuals, porosity between professional and private life, denial practices targeting brands and companies (rumours, hoaxes, chain letters), fragility of information systems (viruses, Trojan horses, worms, denial of service, spam, phraming, phising, hacking, etc.), defamation, incompetence and digital illiteracy, information overload, etc." (Papy and Sansonetti 2014, pp. 21–22).

Despite these contradictory views, the presence of ICTs has continued to grow in organizations, until recently leading to a form of digital hybridization.

1.4.1. *Digital hybridization of organizations*

Regardless of their sector of activity, organizations are making digital an essential part of their business to the point where it has come to condition the overall functioning of the organization in the same way as each structural component. The penetration of digital technology in all operational and administrative services produces a digital artifactualization of the organization, which tends to hybridize: digital technology becomes intertwined with its core business to create a hybrid organization (Battilana and Lee 2014; Huang *et al.* 2017). The organization must integrate different operating logics (Pache and Santos 2013) such as decentralization or the duality between the historical activity and the development of new services (personal spaces on the site, online shop/physical shop for industrialists, internal and external communication). Organizations thus develop both offline activity (linked to the historical heart of the organization) and online activity (such as new services, personalized spaces, etc.) with the need to make them work in synergy.

This technical hybridization manifests itself in the development of specific solutions (intranet, ERP, internal digital social network, electronic document management) (Wihlborg and Söderholm 2013; Winter *et al.* 2014; Bouillon 2015). Subsequently, these new devices are accompanied by "new forms of 'hybrid' collective action inspired by the 2.0 organization, but reinforcing pre-existing values, project and structures of meaning, partly induced by national cultural frameworks" (Dudezert *et al.* 2015, p. 39). The organization is being shaken by the convergence of the technical and the social. While the technical contributes to structuring practice, the social innervates it by influencing uses and practices (Jouët 1993).

Digital communication tools disseminated internally by organizations play a central role in the coordination of participants in collective action (Borge-Holthoefer *et al.* 2012) even if the modalities of this action do not depend exclusively on technology (e.g. if relational networks are pre-existing, they are equipped with software such as instant messaging). While it is certain that virtual communities (see section 3.4.2) could not have emerged without ICTs (and the lower costs of coordination or participation in collective action induced by their use (Lupia and Sin 2003)), digital technologies are a source of tension and risk for individuals that are "prisoners of the means they have used to regulate their cooperation and which circumscribe even their ability to define new purposes" (Crozier and Friedberg 1977, p. 17).

1.4.2. *Digital uses: tensions between private and professional life*

Indeed, this hybridization is accompanied by an increasing dissemination of uses between the public sphere (in the media, within the company, etc.) and the private sphere (personal framework of the life of individuals, family or friendship circles, etc.) (Rodota 2011). The use of tools or services (such as web portals, file transfer services) in both professional and personal settings is becoming increasingly similar. Thus, the Bring Your Own Device (BYOD) trend is indicative of this private/public convergence through the transformation of personal equipment into work equipment.

This seems to be becoming a *de facto* standard. Within organizations, this state leads to situations between the individual and the organization that generate several types of friction:

– Pressure from employees, as mentioned above, with their rich skills and knowledge that they apply in their personal lives. This can lead to conflict (or frustration) between employees and management teams on governance issues (Davison and Ou 2017). This combination of circumstances then leads to the rise of the shadow IT phenomenon (Silic and Back 2014; Silic *et al.* 2016) or the use of unauthorized[31] online tools.

– Tension between, on the one hand, control by the organization and, on the other hand, the promise of ICTs as carriers of "autonomy, cognitive power, relation activities" (Granjon 2009, p. 20), notions that individuals tend to implement both in their private and professional lives. Thus, setting up a corporate digital social network like Yammer[32] (working with a news wall) mobilizes the same skills for its use (publishing a message, opening a discussion window) as other consumer digital social networks like Facebook or Twitter.

With ICTs, a distancing is taking place with a shift from computing (and its technical dimension) to focus on the social (especially its appropriation and use). This statement echoes the evolution of practices, as well as a recurring phenomenon in organizations: the misuse of technology, also known as "catachresis".

1.4.3. *Diversion of technologies*

When devices are not designed on the basis of the prescribed uses, but nevertheless participate in the functioning of the organization, diversions of use are observed (Perriault 2002; Patrascu 2010). The user finds himself/herself in a situation of discomfort, or even opposition, since he/she has to deal with a system and procedures induced "not so – or not only, it can be said – to bend the tasks to the personal convenience of the employees, but to achieve the objectives that the organization must achieve" (Mallard 2011, p. 13).

The hijacking also manifests itself through the "reflexive uses" (Egnongo 2006) that develop through the use of the device and lead to uses that were not intended when the device was originally designed. For Perriault (2015), this means keeping the technical device, but changing its purpose or the use

31 Their use is not allowed for various reasons, in particular to preserve the confidentiality of an organization's data as well as not to be in violation of the General Terms of Use (GTC) of an online service that restricts free accounts to personal use.

32 Yammer is a corporate digital social network created in 2008 and acquired by Microsoft in 2012. Since then, Yammer has been provided to businesses subscribing to Office 365 packages.

initially imagined. In this logic, Akrich (1998) distinguishes four scenarios, depending on the distance between the initial use and the final use:

– displacement: slight modifications without questioning the initial use;

– adaptation: peripheral adjustments of the device to the needs of the users;

– extension: enrichment of the device by new functionalities;

– diversion: use unrelated to the original purpose.

Metzger (2011) offers another reading and notes that employee circumvention practices remain limited initiatives. Moreover, the changes remind individuals that they remain in a situation of submission and dependence on the organization in which they work.

Regardless of their causes, these misappropriations of technologies play a full part in the process of appropriation of the devices by their users. Proulx (2005) proposes four conditions necessary for the appropriation of devices:

– the development of a technical, as well as cognitive, mastery of the device so as not to be only a passive user, but to understand its use and its operation;

– the integration, which must be significant, of the technical object in question into the daily practice of its users so that its use becomes more "natural";

– the repeated use of this technology, which may lead to novelty in the practice;

– taking the system into account at a collective level, so as to achieve a broad appropriation by users and to concretely include the positive device in the organization's policies.

Indeed, users try to personalize them, to fit them into the environment in which they operate. The appropriation process is then essential as a phase of internalization and exploration of the technical device, both in its individual and collective dimensions.

Throughout this chapter, we have placed the organization in the context of the digital age and the growing influence of information in our society. To illustrate this, we have returned to the duality of temporalities between digital and organization and presented the beta logic, very present in the Web environment, which is taking hold in organizations. The introduction of technologies in the organization crystallizes this acceleration of time, the devices being constitutive of the action. Indeed, technical devices, such as information systems, are positioned as mediators of action situations and influence the very functioning of the organization. All this leads to the question of the transformation of organizations that hybridize with ICTs and the uses that result from them. These transformations lead to new reticular organizational forms strongly marked by digital technology, one of whose declinations is the platform.

2

The Platform Organization

The transformations brought about by new technologies are leading to increasingly hybrid organizations, as illustrated in Chapter 1, which must integrate different operating logics, such as decentralization[1] (Huang *et al.* 2017) or the duality between historical analogue activity, and the development of new digital services. Indeed, with the development of ICTs and the transition to the "information society", most organizations are simultaneously carrying out offline and online activities (Prasarnphanich and Gillenson 2003). The information system[2] (Reix 2004; De Vaujany 2009; Mu *et al.* 2015), as a technical device, then takes a central place in accompanying the technological mediation of ICTs and their relations with end-users. After having introduced and defined the concept of platform, we will return to the platformization of organizations with the example of the State considered as a platform to illustrate how it works in concrete terms. Finally, we will conclude on the question of knowledge with regard to the platform and its specificities.

In the framework of this book, we place ourselves in a perspective (Chapters 3 and 4) where the platform can be conceived as a technical and technological base available to the actors of the organization, a base that the latter can mobilize according to his/her needs, in a logic of collaborative platforms (Compain *et al.* 2019), and this is in contrast to the capitalist

---

1 This is favored by the possibilities offered by technical devices such as teleworking.

2 We use as a definition of an information system the following acceptance: "an organized set of resources: hardware, software, personnel, data, procedures… that enables the acquisition, processing and storage of information (in the form of data, text, images, sounds, etc.) within and between organizations" (Reix 2004, p. 3).

platforms that will be discussed in this chapter (AirBnB or Uber, for example), as analyzed by Rosenblat (2018) or Srnicek (2018).

The omnipresence of digital actors in the functioning of a society marked by the orientation of the economy towards the exploitation of information and data tends to the emergence of a new form of organization: the platform organization. It results from the deployment of technical systems and aggregated and collaborative digital marketplaces that organize the supply and demand of services and goods of all kinds on a global scale. Studied and questioned by a broad scientific field (such as computer science, economics, management, law, sociology, information and communication sciences or geography), this digital transformation of market models is giving rise to contrasting reactions.

For some, it is at the root of profound concerns about a re-instantiation of a predatory platform capitalism (evolution of work to task, robotization of human activities). Thus, the sociologist Rosenblat (2018) immersed herself for three years in the world of Uber drivers. She explains how the company operates, the relationships it has with the drivers, and underlines the way the algorithm operates so as to maximize Uber's profit by encouraging drivers and customers to use the application[3]. Throughout her investigative work, she also describes the daily life of the drivers and the precariousness of their situation. In addition, the economist Srnicek (2018) details his vision of platform capitalism. By going back over the genesis of the major platforms and contextualizing them in the capitalist system, he describes the monopolization of data by these actors as an economic asset and the digitization of the neoliberal economy in response to past crises. These examples illustrate concrete aspects inherent to the functioning of the platform, as well as the fundamentals on which it is built, in particular the logic of intermediation and the non-proprietary ownership of the means of production[4].

Conversely, other thinkers see the platform as a perspective for the transformation of our society that serves as a vehicle for economic opportunities and growth (Parker *et al.* 2016; McAfee and Brynjolfsson 2017).

3 For example, to ensure that a driver will be available in a remote area but where an Uber user is located for a hypothetical race.

4 For example, Uber is the largest taxi company in the world without owning a single taxi, as is AirBnB in the hotel sector.

In particular, Parker *et al.* discuss the construction of current platforms and how to create them, while McAfee and Brynjolfsson provide a global picture of the digital transformation that would be at work in our industrialized societies.

In France, this unprecedented transformation of organizations has been reflected in the functioning of the State, which, by initiating the dematerialization of public services[5], has opened up to a form of institutional platformization. It is a complex process of reorganizing public services, designed to make them more accessible while improving their internal functioning, that has led to this institutional platformization. The reconfiguration of government structures in the form of a distributed architecture has led to the establishment of a reticular organization.

2.1. Networking of the organization

ICTs present themselves as "revealing the emergence of new organizational forms" and the Internet plays the central role of coordinating collective action as a "metaphorical device for organizational processes and forms" (Le Moënne 2006, p. 49). Thus, the objective identified, according to Benghozi *et al.* (2002), is linked to the implementation of these technologies in the organization so as to successfully overcome the logic of the hierarchical organization chart in favor of a horizontal organization. As a result, the reticular organization that is being deployed is expressed in the form of a system that makes massive use of ICTs. Communication tools (such as e-mail, instant messaging, mailing lists, forums) and collaborative mechanisms (production and sharing of documentary resources, shared diary, project management) constitute the common functional base. If this change in logic translates into greater freedom of action for actors, does it result in the multiplication of relationships inherent in the reticulation of new forms of control[6]?

5 The administration transformation program, launched in October 2017 and named "Action Publique 2022", is based on the digital transformation of administrations, with the objective of dematerializing all public services by 2022.

6 Notably through self-control (empowerment of teams, for example) or control by the collective.

2.1.1. *Foundations of the reticular organization*

The reticular organization, as it is understood today, is the result of convergences: on the one hand, the construction of large distributed organizations (several thousand employees spread over several cities in the same country, or even in several countries) coupled with globalization, which has fostered the international division of labor, and, on the other hand, the increasing digitization of exchanges and socio-professional relations, facilitating the coordination of dispersed teams. All this in an environment of growing international competition, similarly fostered by globalization and the interconnection of economies.

In this respect, it should be remembered that ICTs, imbued with the libertarian counter-culture that was very present in Silicon Valley in the 1960s and 1970s (Turner 2012), contribute to reinforcing the individualization process that accompanies capitalist thought[7]. Indeed, reticular organization is above all thought and constructed as opposed to centralized organization, which can be associated with administrative red tape. We find characteristics that favor flexibility (reticular organization appears as a dynamic network phenomenon) and the adaptability of organizations, following the example of the attributes identified by Ferrary and Pesqueux (2004):

– *plasticity*: inherent to its nature, the network has a capacity to integrate/disconnect actors, contents, ramifications, without destabilizing the functioning of the whole (see Figure 2.1 for an analogy with the ancestor of the Internet: Arpanet);

– *transversality*: the logic of the network makes it possible to free oneself from spatio-temporal constraints or social determinisms opening up possibilities for diverse actors to meet. It then has an integrating dimension;

– *selectivity*: each actor has the possibility to choose his/her channels to communicate, to disseminate/receive information. This possibility can lead to more complexity by multiplying the communication channels, which in

7 The symptomatic concept of this thinking being methodological individualism, technologies put the emphasis on the individual, in particular on the power given to everyone by these technical devices. Indeed, the said technologies are regularly associated with the English-language term *empowerment,* which underlines the capacity of individuals to empower themselves to develop their power to act.

today's organization is reflected in the increasing number of communication devices such as instant messaging, forums or digital social networks.

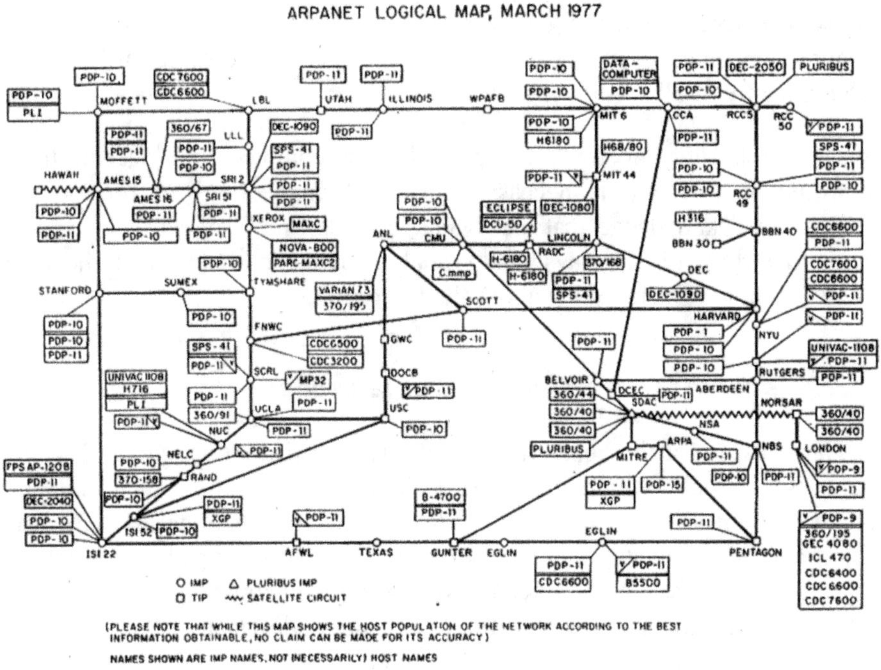

Figure 2.1. *Representation of the Arpanet network in March 1977 (source: Computer Museum)*

COMMENT ON FIGURE 2.1.– *The map shows the beginnings of the Internet, including how the network works. On the Arpanet map, we find the major American universities such as Stanford or Hawaii, as well as the connection of the universities with the Pentagon. It is this same logic which is still present at the present time although it applies on a planetary scale. Any user with a compatible device is then able to connect to the network and access the content made available. The ramification of organizations in this reticular perspective favors the integration of this connection/disconnection logic within their very functioning.*

The reticular organization thus obeys a logic of connectedness[8] and optimization. It holds out the promise of efficient, effective operation and, by

8 That is, all the elements that make up the network are connected and interdependent.

extension, improved social relations (thanks to their mediation by technology). This type of organization would make it possible to overcome the limits, biases and subjectivity of human thought. It is then a structure that seems objective and without judgments or values. In this perspective, the organization is seen as a system that must be repaired if it fails, independently of all social considerations.

Thus, the reticular form is adapted to a complex and decentralized environment where it is necessary to build connections to browse the resources and knowledge that are available in its ramifications. It is for this reason that Morin (2015a) says that the analogy, made in French, with the Web is a good choice when talking about the Internet.

2.1.2. *A cybernetic reading of the organization*

The network, at the heart of this specific organization (Figure 2.2), thus proposes a model combining cybernetics and computing for which the need for technologies to support and exploit this organizational design was identified at an early stage. The network organization equipped in this way is seen as an interface of contact between its members, regardless of their geographical location, via technical devices that are designated as information systems. Based on a regulatory logic[9] (Supiot 2005), the organization is able to control the entropy it generates in order to continue to function "correctly". It then mobilizes feedback to maintain a homeostatic state.

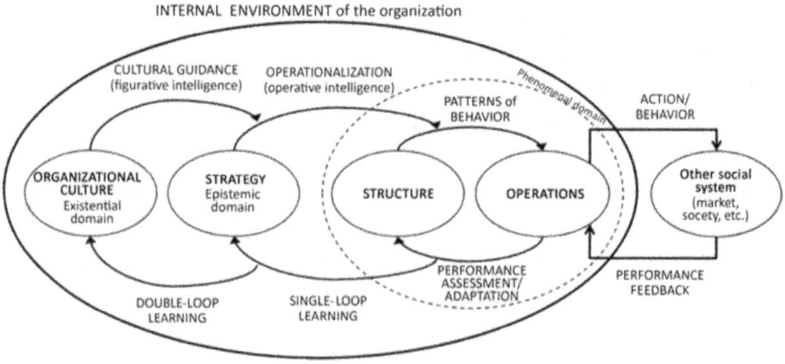

Figure 2.2. *Representation of an organization in cybernetic thinking (source: Organization Orientation Group)*

9 Contrary to a regulatory logic where rules are dictated outside the organization.

COMMENT ON FIGURE 2.2.– *This schematic representation of the organization using a cybernetic design illustrates our point. Cybernetic organization is part of a process of perpetual regulation in order to maintain its homeostatic state. At the same time, based on a principle of continuous improvement, thanks to the implementation of a learning loop, as well as thanks to* feedback, *the organization is more robust.*

This approach is strongly marked by management processes and the associated informatics (such as Enterprise Resource Planning type software packages): "Computerizing means organizing: not only organizing information circles, but also regulating human activity" (Pavé 1989, p. 98).

The generalization of ICTs in organizations promises a rationalization of operations: "Engineers are the experts and heralds of the new configuration of the 'scientific management' of enterprises, namely the network organization" (Ferrary and Pesqueux 2004, p. 27). The quintessence of this vision is today crystallized in what is known as the "platform organization".

2.2. Platformization of organizations

Some organizations[10] see themselves natively as platforms and advocate for this positioning, which allows them to use this legal argument to avoid paying the self-employed who work through their ecosystem[11]. The current trend encourages more and more "classic" organizations to consider themselves as a platform and thus to platform themselves. After a short review of the history of the platform concept and the specificity of associated capitalism, we will illustrate the platformization of organizations with the concrete case of the State.

10 The most frequently cited platforms are those of GAFAM.

11 Their legal arguments are based on the positioning of the company as a provider of an IT service that favors the establishment of a relationship between two independent players who can then carry out a transaction in a logic of supply and demand. This argument is increasingly challenged by various courts such as the Court of Justice of the European Union which, in 2017, stated that Uber's reservation service falls within the field of transport and should therefore be subject to the relevant legislation: https://curia.europa.eu/.jcms/upload/docs/application/pdf/2018-04/cp180039en.pdf.

2.2.1. *Brief history of the platform*

The first publication to explicitly cite the platform is that of Ciborra (1996), which introduces the concept of platform organization in its text, based on a case study, that of Olivetti, in order to highlight its transformation, marked in particular by the influence of technologies that appear over a very short period of time (see section 1.2). The author then mobilizes the concept of the platform that emerged in the 1980s and 1990s in the computer world. Thus, three major currents related to platforms can be identified:

– first, work detailing the establishment of networks of companies in the image of firms or multinationals;

– in parallel, Kerckhove's (2000) research on intelligence networking is a continuation of McLuhan's work on the "global village" (McLuhan and Powers 1989). In addition, Musso's (1998) thinking helps to understand the notion of network, its history and its current declination;

– a third approach focuses on the economy, in particular the work of Rochet and Tirole[12] (2003), who formulate an economic model of the platform based on its positioning as an intermediary between two or a large number of players (two-sided markets or multi-sided platforms). Its link with the market, as a mediator between the producer and the consumer, allows this concept to be inscribed in the logic of the Capitalocene (Morre 2015) or in that of Morozov (2015) for whom the platforms from Silicon Valley are a new emanation of capitalism seeking to ensure its sustainability.

Thus, the platform has evolved to gradually converge towards its current acceptance as a *digital infrastructure that allows two or more players to interact*. The structure put in place is based on a system divided between central components that evolve in a limited way[13] and a set of various peripheral components that contribute to the main objective of the digital platform, which is to keep users in digital or hybrid (digital/physical) spaces controlled by the platform in order to collect as much data as possible. An asymmetrical relationship is established between users, the actors operating through the platform (such as advertisers, customers or even Uber drivers) and the platform itself.

12 Orthodox economist who in 2014 received the Swedish Bank Prize in Economics in memory of Alfred Nobel.

13 Just like the system on which Google's search engine is based and all the associated services developed *a posteriori* by the company Alphabet.

The rise of the "platform organization" is correlated with the development of ICTs, which promotes the emergence of virtual organizations as "a group of network-based systems that can simulate the structure and behaviour of the real domain organisations, and quickly and actively exploit fast-changing business opportunities" (Zhuge *et al.* 2002, p. 325). The characteristics of a platform organization are as follows:

– *autonomous management*: the organization is able to operate on the basis of predefined tasks or business rules;

– *active behavior of members* who are able to perform their tasks freely;

– *intuitiveness of* the virtual organization and its functioning;

– *adaptability and agility* of the organization to quickly adapt to changes in its ecosystem. This acceptance implies in particular for the organization to open up to the other actors with whom it evolves.

These characteristics include two of the fundamental characteristics of contemporary society according to Flichy (2004): autonomy and network functioning. The networked organization also relies on the cross-use of data, which is becoming a major issue for organizations and states.

2.2.2. *The specificity of platform capitalism: culture and use of data*

For Manokha (2018), more than surveillance capitalism, the current trend is towards a platform capitalism where surveillance is fundamental to its ability to create the asset that the platform needs: data. He then proposes a parallel between, on the one hand, Polanyi's thought in *The Great Transformation*, in which labor, land and money have been transformed into "fictitious commodities" and, on the other hand, data, which today would be a new "fictitious commodity" on which platform capitalism and the platforms themselves are based. It is through the use of this data that economic actors grow and produce value. It is therefore vital for these entities to gather data by all means. This explains the recurrence of "scandals" – related to Facebook, Google and other platforms – in the use of personal data.

The major platforms are all the more invasive of privacy because they are based on an English-language conception of the law, known as common law, which is more permissive than Roman law in certain aspects[14]. The debate surrounding the processing of European citizens' data is then emblematic of this issue with the invalidation on October 6, 2015 by the Court of Justice of the European Union of the *Safe Harbor,* the text which provided a framework for the transfer of personal data of European citizens to the United States. In its decision, the Court stated that "the transfer of data of European Facebook subscribers to the United States should be suspended on the grounds that the United States does not offer an adequate level of protection for personal data"[15]. The issue of data then becomes a strong geopolitical and geostrategic issue between states at a time when its quantity is growing exponentially.

Indeed, ICTs are helping to produce more data that are digital traces of reality. These traces have multiplied and their exploitation is facilitated by software that helps to reinforce the use of data (Blouin *et al.* 2009; Bastin and Francony 2016). This state can even lead to a quasi-automatic piloting of the organization which is then *data driven* (Patil and Mason 2015)[16], which completes the "organization-process" of Bouillon (2015), the data being used to orient the organization according to its inputs. This paradigm shift from information culture to data culture results in a shift from a constructivist approach, where information is constructed, allows for the unexpected and is "closely associated with uncertainty" (Gleick 2015, p. 203), to a neo-positivist thinking that is used to justify choices and decisions via algorithms that are biased by nature and by the way they are implemented (Stiegler 2015; Calude and Longo 2016).

But how do we define this data culture? Authors such as Illien *et al.* (2013), Plantin and Valentin (2013), Rousseau *et al.* (2014), Collet-Thireau and Thomas (2015) or Patil and Mason (2015) all evoke the "data culture", but without ever defining it or providing its issues and problems. Nevertheless, Collet-Thireau and Thomas propose a list of skills that mixes knowledge and know-how: "computer and technical skills: understanding the organization of digital [...] and information systems that enable dialogue with computer scientists; being able to disseminate data via a portal,

14 See Supiot (2005) for a better understanding of the differences between Roman law and common law.

15 The decision is available on the website of the Court of Justice of the European Union: http://curia.europa.eu/juris/documents.jsf?num=C-362/14.

16 A data-driven organization is then driven by data in decision-making situations in particular.

facilitating access to and use of data, downloading; competence in data processing [...]; legal competence: knowing the laws governing the use of data [...]; managerial competence: knowing how to manage and lead a project; reasoning in an ecosystem, working in a team, developing a 'data culture' among collaborators, monitoring the use of data" (Collet-Thireau and Thomas 2015, p. 10).

This list is not fully satisfactory since 1) it mixes knowledge and know-how intended for both managers and more technically advanced teams and 2) it is recursive in emphasizing that to develop a data culture, it is necessary to develop a data culture among employees. It is, however, richer than the description by Illien *et al.* (2013), who limit data culture to IT skills related to "business" knowledge. Patil and Mason (2015) propose a draft answer to this question by working mainly on data-driven organizations such as Google, Amazon, Facebook or Apple. They present what is assumed to be for them the expression of the data culture: the presence of *data scientists*[17] and the implementation of a data-driven organization, which should lead to greater efficiency or cost reductions. Patil and Mason finally propose to democratize the data by making it accessible in the organization but by providing these various elements; they do not really define what the data culture is.

In the case of Blouin *et al.* (2009), Lloveria (2014) or Zambon and Monciardini (2015), they focus in particular on data processing, visualization and analysis needs but not on meaning. According to Bachimont (2011) and Gleick (2015), this question of meaning, acquired in particular through context, remains an essential issue. For an organization, it is also an important issue in terms of constructing common meanings and meaning for its members. Indeed, Hachour (2011) highlights that the co-production of meaning is a driving force of the organization; here, it is a co-production of meaning based on the organization's intangible heritage and thus makes it possible to "construct new common meanings" (Guez 2007, p. 98).

We can then propose the following definition for the data culture: *all the knowledge, know-how and interpersonal skills enabling to identify, process, analyze, link data in order to be able to use the data available to solve a*

17 Literally, "data scientists" who use a combination of mathematics, statistics and programming knowledge in their work and who are good communicators (Peng and Matsui 2015; Shan *et al.* 2015; Blum *et al.* 2016).

problem or to obtain an answer. It is then a form of expression of algorithmics and mathematization of the world by digital technology.

This new data culture is driven by the implementation of a "data economy" promoted by both institutional (Lacombe *et al.* 2011) and private actors. In particular, they convey the myth that the data made available "can be reused by the community of developers and entrepreneurs, who can invent new uses for it" (Lacombe *et al.* 2011, p. 22).

The data culture thus contributes to the myth of progress, which today is centered on innovation, by suggesting that more and more data make it possible to answer more and more questions (Bouveresse 2002). This approach to data culture is the bearer of a particular model of organization (data-driven) strongly marked by IT and management (in the sense that it wants to control the whole). We arrive at highly rationalized organizations – they are small and limited – where "the organization of men's work will eventually have to be translated into binary language" (Pavé 1989, p. 238). It is then a functionalist and managerial organization, but one that is less innovative and goes in the direction of the "automatic society" described by Stiegler (2015).

Companies that are rooted in this platform capitalism adopt a capitalization coupled with a massive use of personal data and data generated by the multiple *ad hoc* arrangements. This trend accompanies the de-industrialization of rich countries[18] towards the "knowledge society".

UNESCO has been working on "knowledge"[19] since 2005, where the value of knowledge is produced through cooperation in the form of a "cognitariat" of "knowledge workers".

18 For example, the tertiary sector, defined by INSEE as "a vast field of activities ranging from commerce to administration, including transport, financial and real estate activities, business and personal services, education, health and social work", is strongly marked by this dematerialization and represents in France approximately 80% of gross domestic product (GDP).

19 This society rests on four pillars: freedom of expression, universal access to information and knowledge, respect for cultural and linguistic diversity and quality education for all. UNESCO then posted on its website that "the creation of knowledge societies is based on the conviction that universal access to information is essential for the construction of peace, sustainable economic development and intercultural dialogue. UNESCO promotes "'transparency' in content, technology and processes through awareness-raising, policy development and capacity-building" (https://en.unesco.org/themes/building-knowledge-societies).

2.2.3. *Towards the construction of a "platform organization"*

An exacerbated form of reticular organization, the "platform organization" is comparable to a "biological superorganism" (Marciniak 2013). The global entity is based on symbiotic interactions between the different actors of the organization who have a certain autonomy. In order to function, this organism mobilizes several means: ICTs to organize the interactions of the entities and organizational means based in particular on self-control and motivation of the personnel involved.

What then are the distinctive features of this new organization? For Marciniak, the organization is based on:

– The permeability of the boundaries between entities, which results in varying degrees of control over the entry or exit of an actor from the system and its contributions. This structure tends to favor the emergence and development of communities and/or collectives that contribute to achieving the defined objectives. It is this same permeability that we already found in the discourse around the network (Ferrary and Pesqueux 2004).

– Common objectives, directly or indirectly, which bring together the actors who mobilize skills and knowledge in return. These actors evolve in a context that combines infrastructures (information technology, for example), communication and collaboration protocols coupled with processes (decision-making, in particular) and common resources (knowledge, budgets, etc.). The value generated by all the interdependent actors is shared among all to ensure the sustainability of each and their interest in becoming involved in the organization. These elements contribute to an environment of trust and mutual aid necessary for the accomplishment of missions.

– The members of this super-organism have a clear vision of what is controlled and by whom within this ecosystem where they are interdependent, contributing to the formation of a common identity (Ahrne and Brunsson 2005). This vision is essential for anticipating difficulties and conflicts between the parties involved in operational decisions (Ahrne and Brunsson 2005). This aspect should not be underestimated given the importance of the autonomy that individuals gain and exercise in this type of organization. The motivation that manifests itself in their potential to contribute is certainly a source of creativity, as well as of tension when competing proposals emerge. They then require sensitive and subtly balanced trade-offs (Gulati *et al.* 2012).

– The business model is defined by the distribution of value among members, as well as by the way in which value will be created and capitalized.

The characteristics mentioned are expressed to a greater or lesser extent depending on the organization. Indeed, the diversity of the data produced and the increasing digitization of activities contribute to the creation of many types of platforms – sales (such as Amazon), services (such as Facebook or Google) or cooperatives (such as those present on the Coop des Communs[20] website, which brings together cooperative actors present in this form) – which can be very different from one another. This organizational orientation does not spare the functioning of the State which, engaged in a dematerialization process, gives shape to a more or less successful expression of public service platformization.

2.3. The platform state

While platform operation originated in the private sector, its adaptation to the state and public services is becoming more widespread (Janssen and Estevez 2013). In France, the State's platformization doctrine is driven by Henri Verdier, ambassador for digital technology and former interministerial director of digital technology and the State's information system. He detailed his vision in a book published in 2012 (Verdier and Colin 2012) and developed it in a study conducted in 2017 with Fondapol[21] on the strategic orientations for the State's information systems. His recommendations aim to set up "state start-ups" within the ministries. These state start-ups are, in fact, perceived as light and reactive institutional arrangements to improve the lives of citizens by mediating their relations with the administration.

In the text by Verdier and Pezziardi (2017), we find a thought close to cybernetics which consists of "innovating to repair the system". The State appears as an organization crossed by flows, which is based on a system that must improve on a daily basis to optimize its performance and go as far as the automation of tasks and processes.

20 https://coopdescommuns.org/fr/plateformes-en-communs/.

21 The Foundation for Policy Innovation is a think tank that presents itself as liberal, progressive and European: http://www.fondapol.org/etude/pierre-pezziardi-et-henri-verdier-des-startups-detat-a-letat-plateforme/.

FranceConnect (Alauzen 2019), based on a Single Sign-On (SSO) authentication system, illustrates the platformization of public services in France as an extension of the strategic orientations of Facebook, Google and Twitter[22] platformization. Verdier thus revisits the State in the light of platformization. This current of thought is strongly linked to the logic of open government[23] through the liberal vision of the State and in particular by refocusing on its regalian functions.

2.3.1. *The thinking of the State as a platform*

This political thinking of the State is part of a trend correlated with capitalism and libertarianism, which translates into a distrust of the State and an exclusive refocusing of the State on its regalian functions alone. In 2010, the Gov 2.0 Summit congress was held, which introduced the concept of Government as a Platform to the world. The stated goal of the Government as a Platform[24] was to "bring together innovators from government and the private sector to highlight technology and ideas that can be applied to the nation's great challenges. In areas as diverse as education, health care, energy, employment and financial reform, there are unique opportunities to rethink how government agencies perform their mission and serve our citizens. Social media, cloud computing, Web and mobile technologies all provide new capabilities that government agencies are beginning to harness to achieve demonstrably better results at lower cost". These stated objectives thus lead to considering private sector action in the missions carried out by the state and fighting bureaucracy[25]. Tim O'Reilly, the essayist behind the concept of Web 2.0, describes the "platform state" in his 2010 book *Open Government*: "Government 2.0, then, is the use of technology – especially the

---

22 Which allow you to connect to sites directly with your Facebook, Google or Twitter account. The interest of these actors is to retrieve their users' data on other sites and to continue to study their online behavior.

23 In 2011, several States and civil society organizations have joined together to launch the Open Government Partnership, which campaigns for transparency in public action and its openness to new forms of consultation and collaboration with civil society, in particular by leveraging digital and new technologies: https://www.opengovpartnership.org/.

24 Government as a Platform is a term that echoes the business model of Software as a Service (SaaS). This proximity tells us something about the logic behind government.

25 Thus, the title of Andrew McAfee's speech is quite explicit: "Gov 2.0 vs. the Beast of Bureaucracy". The text can be consulted at this address: https://web.archive.org/web/20160323040737/http://andrewmcafee.org/2010/09/oreilly_gov20_summit_bureaucracy/.

collaborative technologies at the heart of Web 2.0 – to better solve collective problems at a city, state, national and international level" (O'Reilly 2010). Technology, particularly the network, is then a solution to the problems encountered by states and public administrations due to the lack of rationalization of their actions. We find an optimization objective, in fact, the result should lead to a lean[26] government, i.e. moving away from the Leviathan figure (Hobbes 2000) to do more with less (notably less means in a liberal vision) (see Figure 2.3).

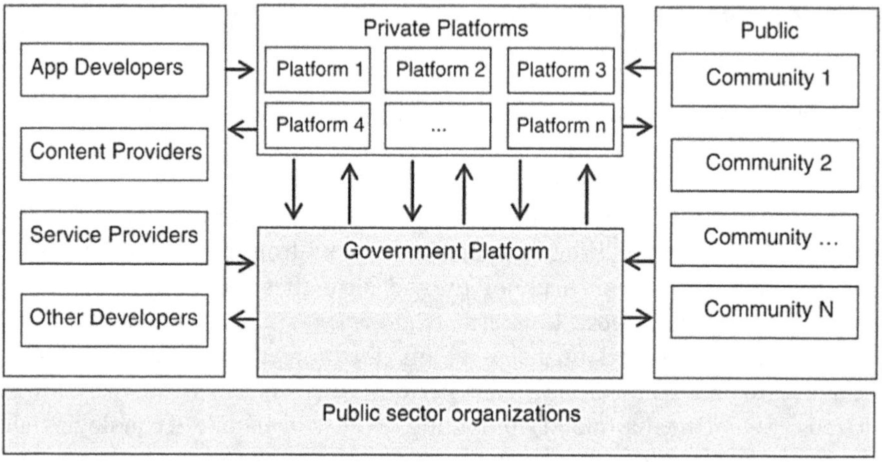

Figure 2.3. *Representation of the platform state (according to Janssen and Estevez (2013, p. 3))*

COMMENT ON FIGURE 2.3.– *This representation of the State as a platform places it in an ecosystem where it provides a service to the various stakeholders using data, whether they are citizens, private platforms or digital service developers/providers. The State is then one actor among others. In their article, Janssen and Estevez thus argue for the transformation from*

26 The term *lean* refers to the implementation of a *lean* production system to avoid waste. It is part of a tradition of performance research designed to reduce waste, which can be of several kinds: overproduction, waiting, transport, unnecessary steps, stocks, unnecessary movements, corrections/retouching. While this logic is rooted in the system set up by a private company, Toyota, it tends to apply to States. Thus, in France, lean was present in the 2007 general review of public policies (RGPP) (see the review of the RGPP and the modernization of public action published by France Stratégie: https://fichiers.acteurspublics.com/redac/pdf/2018/2018-03-07_FRANCE-STRATEGY_Balance-RGPP-MAP.pdf).

an omnipresent State to a smaller State, accompanying citizens and private actors to respond to societal challenges.

Parallel to the "platform state", the doctrine of open government or open democracy was constructed.

This doctrine is based on the presuppositions of improving the efficiency of governance and transparency.

It must then promote citizen participation and collaboration between the different strata of society.

The dimensions covered by the establishment of an open democracy (or open government) logic are summarized in Figure 2.4.

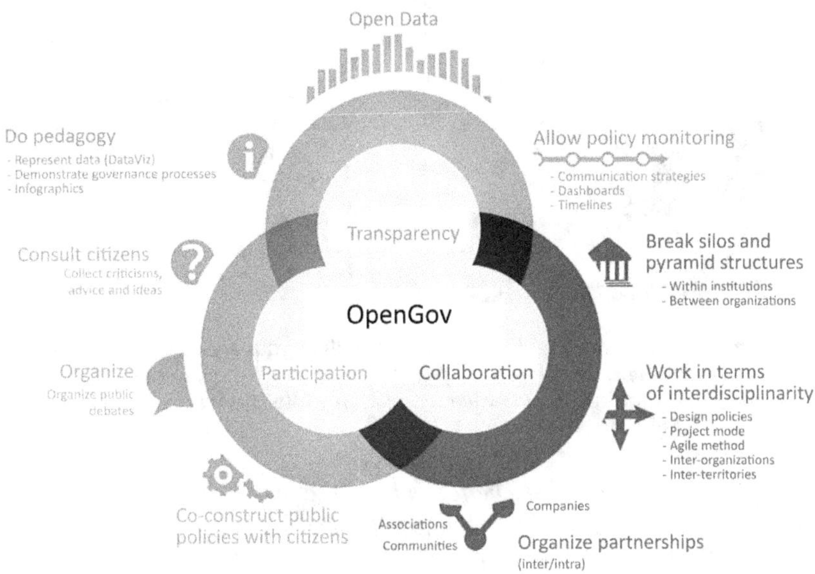

Figure 2.4. *Illustration of the set of areas that are addressed by the government in the framework of open democracy (source: Armel le Coz and Cyril Lage, CC-BY). For a color version of the figure, see www.iste.co.uk/henry/intelligence.zip*

COMMENT ON FIGURE 2.4.– *In this infographic, the fundamentals of open government are transparency, participation and collaboration. The authors explain these major axes with operational variations (such as the opening up of public data) or guidelines (such as the implementation of co-constructed or transversal approaches). This schematic representation expresses the transformation of the state apparatus into as many start-ups with a desire to break down the silos.*

2.3.2. *The State, a platform like any other?*

How is the State as a platform? On the website modernisation.gouv.fr (Figure 2.5), there are available documents and articles describing this form of state.

Figure 2.5. *Screenshot of the modernisation.gouv.fr site on the page presenting the State as a platform. For a color version of the figure, see www.iste.co.uk/henry/intelligence.zip*

COMMENT ON FIGURE 2.5.– *The focus is on the production of public services using digital technology. These digital services must be simple for all audiences. However, this issue of inclusion of all audiences is far from being completely resolved at present. Illectronism*[27] *and the Digital Divide (Le Guel 2004; Bacache-Beauvallet et al. 2011) are sources of exclusion from citizenship. In his annual report for 2018, the Defender of Rights highlights the consequences of dematerialization in terms of inequalities in access to*

27 The INSEE survey on illectronism sheds light on this social reality: http://www.epsilon.insee.fr/jspui/bitstream/1/110065/1/ip1780.pdf.

public services[28]. *In particular, it refers to the "Public Action 2022" plan, which gives priority to the digital transformation of administrations, with the objective of dematerializing all public services by 2022.*

Related to this example, the beta.gouv.fr site (Figure 2.6) is also significant in relation to the logic that is being put in place. This site shows the digital ecosystem in which the platform state is embedded.

Figure 2.6. *Screenshot of the beta.gouv.fr site. For a color version of the figure, see www.iste.co.uk/henry/intelligence.zip*

COMMENT ON FIGURE 2.6.– *The State is part of this permanent prototypical logic with State start-ups. On the website beta.gouv.fr, services under development are accessible. The use of the term "beta" represents an important change, as noted by Verdier and Pezziardi (2017, p. 22): "A site in .gouv.fr displays a 'beta' banner. This is a low symbolic posture, synonymous with work in progress and continuous improvement". Prototyped digital services are developed using an agile method, with the result that they are put into production regularly and are confronted as quickly as possible with users so that they can "pivot" (the emblematic term for this logic) quickly if necessary. The main principles evoked on the page (listening to user needs, continuous improvement, steering by purpose) are those of agility.*

Both the beta.gouv.fr site and the projects presented online are accessible on the GitHub[29] code sharing site.

28 The report is available on the Advocate's website: https://www.defenseurdesdroits.en/sites/default/files/atoms/demat-num-21.12.18.pdf.

29 https://github.com/betagouv.

This sharing of source code would contribute to the transparency of government action. Internet users (citizens) can contribute to improving the source code of tools developed by the State by auditing them to suggest improvements.

However, this desire for democracy presupposes technical skills and having the necessary teams to participate effectively in this digital democracy.

Figure 2.7. *Screenshot of the api.gouv.fr site listing the APIs made available by the government. For a color version of the figure, see www.iste.co.uk/henry/intelligence.zip*

COMMENT ON FIGURE 2.7.– *Each API is described and sources the origin of the data that is accessible. Documentation is systematically provided to enable the use of the API as for all global enterprise marketplaces.*

Finally, to complete the offer of the French platform-state, the api.gouv.fr website lists the programming interfaces of state applications (Application Programming Interfaces, APIs), which are computer services accessible from the Internet to carry out specific operations (Figure 2.7). The implementation of APIs open to all then echoes the representation of Janssen and Estevez. Indeed, the arrows connecting the stakeholders are most often APIs, in the context of IT interactions. The State becomes a *de facto* provider of online services (Figure 2.7). The first beneficiaries are both administrations that request data held by other administrations and businesses that can integrate these services into their offers or tools. While we find similarities with the major commercial platforms that also make beta services available to APIs, the government stands out for its very essence and the non-market dimension of its activity. The emblematic example, and correlated with the idea of *opengov,* is that of the opening of public data (open data). The State, by legislating[30], imposes several obligations[31] on administrations as well as on public service providers in connection with the information society and open democracy.

However, this obligation to be open leads to questions about the beneficiaries and the model for sharing this data. The questioning of the repercussions of this process of opening up public data was raised in particular by Goëta and Mabi, who point out that "the hope of the discovery of a 'new oil' has long dominated the debate on open data, relegating to the background its citizen issues in terms of transparency and emancipation of citizens" (Goëta and Mabi 2014, p. 83), which is in line with the evolution of the political project towards the economic project of the "information society". Moreover, researchers note that "data are often selected and reworked to reduce the complexity and risks of their use by an unknown public" (Goëta and Mabi 2014, p. 85). The fears expressed by these authors are corroborated by the work of other scientists who are working on resolving the tensions between the ethical (by promoting participation in democratic life, for example) and economic dimensions of data openness (with a form of subcontracting being put in place) (Sieber and Johnson 2015). Thus, for the time being, empirical research on data openness fails to demonstrate its role in the democratic process (Ruijer *et al.* 2017).

The example of the openness of public data nuances the very positive vision of the platform state held by different governments. Indeed, the establishment of the platform should, according to the advocates of this vision, make it possible to respond more effectively and efficiently to the

30 In particular with the law no. 2016-1321 of October 7, 2016 for a digital Republic, all the files of which are accessible on the Légifrance website: https://www.legifrance.gouv.fr/afficheLoiPubliee.do?idDocument=JORFDOLE000031589829&type=general.

31 The website https://www.economie.gouv.fr/republique-numerique lists the essential points of the law and in particular the obligations such as the opening by default of public administration data (accessible on https://www.data.gouv.fr/), the opening of energy consumption data (for energy actors, accessible on https://opendata.agenceore.fr/) or the free exchange of data between State administrations.

needs of citizens. Nevertheless, tensions and contradictions emerge between a technical-driven ideal and the reality of usage.

To complete this question of the platform, we will address the knowledge angle and more particularly knowledge organization within this type of entity.

2.4. Platform and knowledge

The link between this new organizational form and knowledge organization is close; the platform, decentralized by nature, is natively based on hybrid systems of knowledge organization that are information systems and is part of this "knowledge society" promoted by UNESCO.

Digital technology, marked by the value of sharing (Saleh and Hachour 2012; Verlaet 2015), also seems to be a tool facilitating the dissemination of knowledge to the greatest number. Indeed, whether it is through Web 2.0 (or Social Web), influenced by the founders of computer science and libertarian ideology, the establishment of the sharing economy or collaborative economy (Acquier *et al.* 2017; Mair and Reischauer 2017), collaborative tools (Lyytinen and Ngwenyama 1992; Olson and Olson 2008; Silva and Ali 2010), digital is converging towards sharing.

There are a number of benefits that facilitate the sharing of digital content, in particular the cost of sharing. Indeed, the marginal financial cost of reusing data is close to zero, since the use of a digital resource does not deprive another person of its use; this is what the principle of non-rivalry in economics teaches (Samuelson 1954). However, without a willingness to share knowledge, fostered in particular by the culture of the organization (Orlikowski 1996a, 1996b), the existence of technical infrastructures will be of little use.

This question of knowledge organization leads the organization to carry out a work by creating its own organizational model and technical devices in order to best meet its objectives by giving the means to act and to say, for Rabardel (2005), "I can" before saying "I know" in relation to knowledge located in an action[32].

32 Argyris (2003) talks about actionable knowledge.

2.4.1. *Knowledge organization within the framework of the platform*

"Knowledge organization (KO) is about describing, representing, filing and organizing documents and document representations as well as subjects and concepts by both humans and by computer programs[33]", and is then to be seen through the prism of the systems that support, produce or participate in knowledge organization, also called "knowledge organization systems" (KOS)[34].

In the case of the platform, it is the question of sharing knowledge internally that is thus addressed, in line with the subtitle of Ballay's work, which states that knowledge is "the only resource that gains value by sharing it" (Ballay 2002). Citing several practical cases[35], Szulanski (1996) emphasizes the difficulty for organizations to share or transfer knowledge. Based on this, how can the members of an organization share knowledge that they are not aware they possess? In his initial observation, Szulanski constructs a reflection around the obstacles to the dissemination of knowledge and thus the reasons that encourage the members of an organization to share or not to share knowledge. Complementing his observation, Ermine indicates that knowledge sharing "is not only understood as a circulation of information, but also as a cross-fertilization between the different actors of knowledge" (Ermine 2003, p. 11).

There are several answers in the literature. To go further on the motivations for sharing, McLure Wasko and Faraj (2000) identify two factors in their work:

– self-interest: may result in financial, reputational, statutory or new functions for the individual. Before knowledge is shared, it is "owned" by

33 According to the definition written by Hjørland in the *ISKO Encyclopedia of Knowledge Organization* produced by the International Society of Knowledge Organization (ISKO): https://www.isko.org/cyclo/knowledge_organization.

34 Which are dealt with exhaustively in the ISKO encyclopedia: https://www.isko.org/. cyclo/kos.

35. The author begins his article by citing the case of General Motors, which is unable to share production practices between group entities, or IBM, whose sharing of software engineering practices is very limited – despite the means made available to ensure effective sharing.

the individual and may be lost to the organization if the individual decides to leave;

– moral obligation: by joining a community, members commit themselves to share their knowledge in a reciprocal way, as well as to keep the community alive and not to be excluded from it. It is then the community that holds the knowledge and is responsible for its dissemination and updating in a collective design.

2.4.2. *Knowledge sharing and dissemination*

Several motivations are present in the organization to implement an approach related to knowledge sharing and dissemination. Mahé *et al.* (2010) identify the following needs: to retain the knowledge essential to the smooth running of the organization, to repair knowledge losses, to make new knowledge emerge in order to do better or anticipate changes or to lose obsolete knowledge in order to help innovation to emerge.

In these contexts of intensive need for knowledge, two types of knowledge emerge.

Identified by Polanyi (1958, 1967) then popularized by Nonaka and Takeuchi (1995), the notions of "tacit knowledge" and "explicit knowledge" are widely taken up by researchers in their work on knowledge (Spender and Marr 2005; Mohamed 2007; Toledo *et al.* 2016).

They are then used to construct representation models and knowledge sharing within organizations like Socialization, Externalization, Combination, Internalization (SECI, Figure 2.8) model (Nonaka and Takeuchi 1995) where tacit knowledge is present in action and is located while explicit knowledge is stored in information systems or archives.

This distinction between tacit and explicit knowledge is similarly found in the two perspectives of apprehension of knowledge developed by McLure Wasko and Faraj (2000): knowledge as object, which then exists independently of human action (cognitivist conception), therefore relates to implicit knowledge. *A contrario*, knowledge embedded in people is from this point of view tacit knowledge that is expressed during a situation of action (constructivist approach). By its very nature, the platform promotes the apprehension of knowledge as an object captured and stored in knowledge organization systems (KOS). In this way, the KOS constitutes a common language, either for the design of a support or, more globally, for exchanges around concerted knowledge created by different supports. It can be used as a framework, a grid to express the shaping of knowledge relating to the field under consideration in as exhaustive and complete a way as possible. Finally, it can be used as a framework to specify the structures implemented in the tools (sometimes simply the outline of a document, but more often the data structures manipulated by a software tool) (Mahé *et al.* 2010).

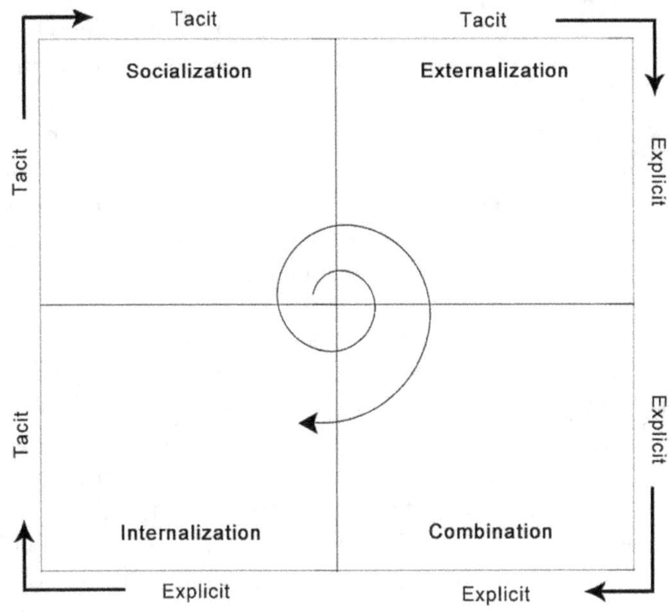

Figure 2.8. *Representation of the SECI model (source: Wikipedia, Ibmgroup, CC BY-SA 3.0)*

COMMENT ON FIGURE 2.8.– *The SECI model models the transition from tacit to explicit knowledge before the knowledge becomes organizational. The first stage of socialization is a phase of sharing tacit (i.e. non-formalized) knowledge in an informal context. The second stage is that of outsourcing, i.e. the formalization of tacit knowledge in documents, training materials, etc. so that the knowledge is accessible. Then comes the combination phase, thought here with the objective of creating new knowledge from existing knowledge. The presupposition is based on the idea that the combination of formalized knowledge favors the emergence of new knowledge that can be assimilated by actors in the fourth (and last) phase, that of internalization.*

Between these two possible conceptions of knowledge, explicit knowledge can be equated with organizational knowledge; it is then "the sum of knowledge in the minds of an organization's employees, and organizational knowledge is increased by people learning or by adding new people to the organization" (McLure Wasko and Faraj 2000, p. 159). Tacit knowledge remains individual knowledge, which members of the organization must share if they wish to transform it into knowledge that is potentially accessible to all. Knowledge can then pass through an intermediate stage, which is the third perspective the authors identify: *knowledge community,* where it is analyzed as "a public good that is socially generated, maintained and exchanged within emerging communities of practice" (McLure Wasko and Faraj 2000, p. 156). In this context, tacit knowledge begins to be made explicit among community members who, according to the authors, have a moral obligation to share it.

Notwithstanding the contextual elements related to knowledge, the problematic of the dissemination of knowledge in the organization also concerns its "sticky" dimension (Szulanski 1996). Indeed, this characteristic increases the cost of sharing according to four characteristics: those of the knowledge transferred, the source, the recipient and the context in which the transfer takes place. Based on this, the author then identifies obstacles to knowledge sharing such as the lack of motivation of the person holding the knowledge or the person who is supposed to receive it, the ambiguity of the knowledge or the lack of confidence of the receiver in the sender. With the identification of these barriers, he emphasizes that the creation of knowledge would then be a "fragile" process that must be accompanied and supported by the members of the organization. He associates the willingness of people to share with *care,* which can be related to concepts such as commitment and the culture of the organization that serve the creation of knowledge by promoting mutual aid.

In this chapter, we have put into perspective the platform organization marked by digital and technocentric capitalism, its functioning, the underlying logics and certain questions inherent to its existence (such as mass data collection). The example of the State as a platform illustrates in particular a vision strongly marked by the technology which must optimize the organization to improve it, to the detriment of the reality of social uses.

With a view to achieving technical and social convergence, Chapter 3 highlights the issue of collective intelligence and its instrumentation in a sociotechnical perspective so as to rethink the platform in an anthropocentric way to serve the members of the organization. This refocusing on internal actors also favors a conception of knowledge linked to the action situations they encounter, moving away from knowledge considered as a "stock" that must be capitalized on in permanent information systems.

3

Sociotechnical Instrumentation of Collective Intelligence

The question of the platform leads us to examine two essential aspects in greater depth: technology (dealt with from a sociotechnical perspective in this chapter) and community (in relation to collective intelligence). The platform, based on a logic of reticular organization, mobilizes these two aspects through the preponderance of technology in its operation and the mediation of the social body by the systems in place, which Guyot summarizes as follows: "the device institutes a specific organization and modes of social relations, even as it uses and creates knowledge and know-how that it converges to produce the tool and its modes of use: choice of the type of training, skills of the project players and of the partners" (Guyot 2000, p. 21). However, the examples of platforms mentioned in Chapter 2 mainly emphasize the technical dimension and the fact that users are subject to this dimension (the example of algorithmic management derives from this logic).

In this chapter, we will first come back to the question of digital literacy before exploring collective intelligence and its possibilities up to the construction of a meta-organization based on this collective intelligence. This path proposes to go beyond the technocentric dimension of platforms to focus on an anthropocentric organization.

3.1. Social instrumentation through digital technology

It is recognized that human activities evolve in contact with information and communication technologies (ICTs):

– in both private and professional daily life (Durampart 2009; Maas 2015): mediation by ICTs weakens to the point of making the borderline between private and professional life (Le Douarin 2007b) invisible. This porosity is confirmed by INSEE figures (see Chapter 1);

– in the uses and practices of information and communication where digital technologies have imposed themselves (Proulx 2005; Comtet 2007; Paquienséguy 2007): digital social networks have increased, for example, the volume of information produced/processed and have led to the quantitative threshold of individuals being exceeded, beyond which communication can no longer take place through an intermediary;

– in the very construction of the organization: this dimension is expressed in several ways, namely, by the implementation of structuring information systems (the famous ERPs) and by the demands of employees who wish to find in their work the same tools or practices as in their private life (Carmes 2010). The massive use of technology in some organizations is transforming them into digital organizations.

While many professional actors (managers, engineers, computer scientists) abound towards a conception where "the technology determines social aspect" (Pavé 1989, p. 10), the fact remains that technological processes are as inadequate for creating communication (Wolton 2000) as they are for building an organization.

Prematurely, Pavé analyzed, in depth, the limits of the technicality of the system by noting that "technologies cannot solve problems that humans, by organizing themselves, do not know, or badly, solve themselves" (Pavé 1989, p. 10). Digital technologies in organization then reveal themselves in the form of "sophisticated technical devices or technical devices that make extensive use of computer communication tools, allowing dispersed teams to work together in project groups" (Cardon 1997, p. 5). For example, information and communication devices using ICTs have been, since the advent of the Internet and its penetration in all strata of companies, massively present in organizations for reasons, proven or not, of increased efficiency and productivity (Jacquinot-Delaunay and Monnoyer 1999; Batazzi-Alexis 2002). Sociotechnics, a specific multi-disciplinary theoretical current, was gradually developed at the end of the 20th Century to study the mechanisms and, in particular, the convergences between technical and social aspects.

Sociotechnical approaches have a common origin, according to Gaglio (2012a, 2012b), in the thinking of Simondon (2014). This "thinker of the technology" paves the way for a "thought of co-construction which allows the question of the relationship between society and technology to be posed differently than by the determinism of one from the other" (Coutan 2015, p. 2). Coutant identifies four major sociotechnical trends:

– In *ethnotechnology*, introduced by Perriault, the focus is on the existing interactions between technology, science and society (Perriault 2008). This current then studies the implementation of behaviors relating to the use of a technology or a device. Perriault follows a logic of use by taking into account five parameters: representation (the user's perception), the social norm of use, the niche of use (role of the object once accepted by society), the construction of a project (where the object, as an instrument, will be used or diverted according to needs) and, finally, the technical footprint (i.e. the influence of the device on its user and the new knowledge produced according to its use).

– The *actor–network theory* is carried out in particular within the Centre for the Sociology of Innovation[1] by three researchers: Madeleine Akrich, Michel Callon and Bruno Latour. These researchers focus on the design of technical objects as results of the discourses and actions of protagonists. The central notion in the actor–network theory is translation, accepted as the set of negotiations, debates, situations of persuasion, etc. that contribute to the action of an actor by offering a capacity of representation where the other is the spokesperson (Akrich *et al.* 1988). By questioning the causes of the success or failure of technologies, this theory contributes to analyzing and tracing contextual elements such as the choices that lead to each situation.

– In *adaptive structure theory*, the authors (DeSanctis and Poole 1994; Brooks *et al.* 2008) claiming to be of this thinking seek to go beyond a neutral view of technology, or one that is considered only in terms of its impact on users. Two major aspects of technology are identified: (1) its *scope*, i.e. its perimeter (such as the physical equipment used by individuals on the job to analyze their interactions) and 2) its role in the organization, which may be subjective, objective or present itself as a "light" technological determinism (Orlikowski 1992, pp. 399–400). The modes of appropriation, the instrumental uses, the freedom of action of the use with the device or

1 www.csi.mines-paristech.fr.

whether the use of the device reflects the ambition of its designer are thus questioned.

– The *double sociotechnical mediation* has been constructed with a view to overcoming technological or sociological determinism. It is twofold by accepting that "this is both technical, because the tool used structures the practice, but mediation is also social, because the motives, the forms of use and the meaning given to the practice are resourced in the social body" (Jouët 1993, p. 101). By offering a possibility of action to the user, the technology becomes concrete and legitimate in social logics, the two mediations are then linked, which leads us to consider both the technical and the social as structuring and situated.

Work on digital instrumentation with a sociotechnical vocation thus highlights the close links, even dependence, between technology and the social context from which it emerges. This research goes beyond the thinking centered on determinism, where two opposing antagonisms exist, according to Fichez and Varga (2009): technologism, where technology determines society, and sociology, which, conversely, considers technology to be socially determined. This antagonism is refuted, in particular, by proponents of the actor–network theory.

3.1.1. *The organization's sociotechnical project*

In the collective imagination, technology has been "a carrier of dreams of immortality, abundance and power, and it has also been, at least since the success of the notion of alienation, the object of a series of criticisms regarding its excess practices" (Martuccelli 2016, p. 37). But currently, these devices that Appel and Heller (2010, p. 43) call "technico-socio-economic devices" leave their place to anthropocentric mediation devices. These devices then carry out sociotechnical mediation with the following characteristics (Wertsch 1991):

– the mediation mechanisms and the actions mediated are socio-culturally situated;

– mediation by the tool always imposes constraints and potentialities (affordances) which can be defined as the actual and perceived properties of the tool and, more particularly, its fundamental properties necessary to determine how the tool is used;

– the device–actor relationship can be characterized either as know-how or by appropriation, both of which are expressed in the relationship between the actor and the device.

While the concept of device still raises many questions (Gavillet 2014), we will accept the following definition, which places it in an infocommunicational context: "a device is an instance, a social place of interaction and cooperation with its own intentions, its material and symbolic functioning, and its own modes of interaction. The economy of a device – its functioning – determined by the intentions is based on the structured organization of material, technological, symbolic and relational means that model, from their own characteristics, the social (affective and relational), cognitive and communicative behaviors and behaviors of the subjects" (Peraya 1999, p. 153). Concerning the devices, two methodological positions are accepted (Appel and Heller 2010):

– research on specific objects, which are then referred to as devices. Three logics are put forward by Appel and Heller: the logic of arrangement (descriptive), the analytical logic (process of social or technical co-construction and meaning) and the critical logic (analysis of effects and issues). Thought as logics of uses, they bring complementary visions of the same object to promote its study;

– searches where "the object is the concept of the device in itself" (Appel and Heller 2010, p. 40), i.e. a work on the epistemology of the device to better understand what is covered by this concept.

We are in the first posture in order to better understand how devices and their related influences work in an organization. The emphasis is then placed on a specific declination of the device – the sociotechnical device. The latter has a strategic dimension by virtue of its role within the organization and its capacity to produce regulation for its users. Undergoing a strong attraction of ICTs (Domenget 2013), sociotechnical devices are based on a translation model focusing on mediation and interactions with the digital world (Baujard 2011b). Their analysis "is a succession of very varied scientific fields by its two technical and human aspects" (Lombardo and Angelini 2013, p. 194) that the currents of sociotechnics propose to study.

Within the framework of sociotechnical devices, we are confronted with social constructions that accompany what Jeanneret calls the "symbolic operativity" of devices, i.e. the representation they produce, the roles, etc., and which are not always easy to understand (Jeanneret 2009, p. 3). For Belton and de Coninck (2007), these devices (and more particularly ICTs) are then, through their mediation role, triple hybrids in organizations:

– they build links or barriers[2] between the different protagonists (individuals, communities and organizations);

– they are socially pre-constructed even though they are specifically designed for each member and their needs;

– they are hybrids by their sociotechnical dimension including technical and social links.

The genesis of a sociotechnical device is correlated with the context for which it is designed, as Akrich (1998) has already identified, and to a problem that it is supposed to address, whether by replacing an old device (such as an IS that is too old for new browsers or with non-functional ergonomics) or to respond to a deficiency (such as a single authentication system). The tool thus produced mediates the activity. It is moreover composed by its material or symbolic dimension or by its modalities of use (Rabardel 1995a, 1995b; Agostinelli 2009), not forgetting that it is "not possible to understand human activity by dissociating humans from tools" (Agostinelli 2009, p. 356).

The device is already part of an artifactual process (Figure 3.1), which begins with the discovery of the object studied, its functionalities, its possibilities before moving towards a possible assimilation of the object in the daily life of the operator and then its appropriation.

Once this stage is reached, the object may even become invisible in its context of use, its "developments are then marginal and the observer tends to consider certain elements of the device as natural frameworks for collective action" (Aggeri 2014, p. 53). However, this vision presupposes effective integration into the uses and practices of users and in their patterns of use, in other words, that the object has become an instrument at the service of the organization's actors. It should be emphasized that the set of objects does not pass all the stages defined by Agostinelli. Indeed, there are some that users do not manage to appropriate and which then remain at a primary stage.

2 Simmel (1988) refers to "bridges" or "gates" in his book.

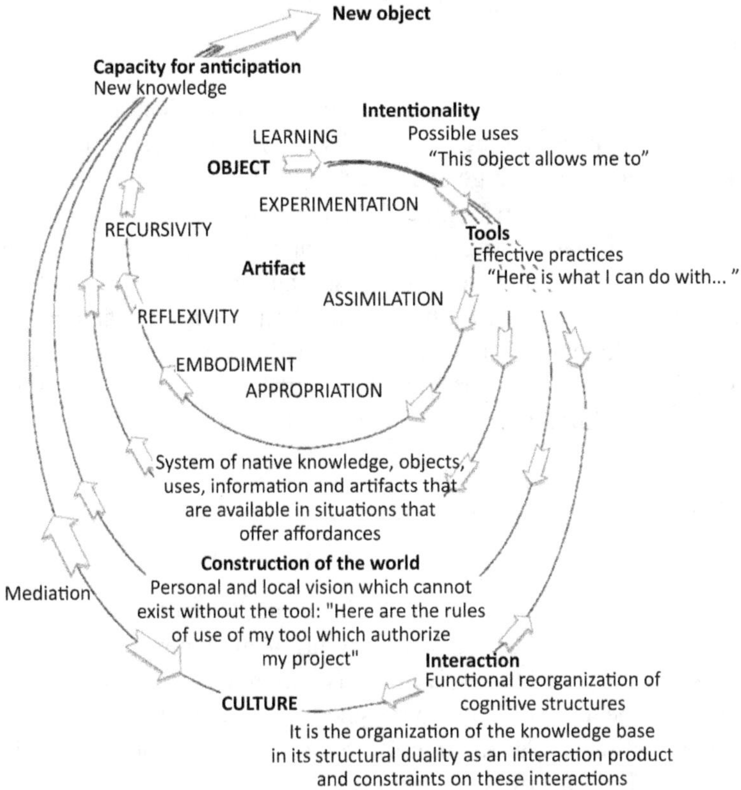

Figure 3.1. *The artifactual process (according to Agostinelli (2009, p. 366))*

COMMENT ON FIGURE 3.1.– *The artifactual process is represented by Agostinelli as follows. It goes through a phase of discovering the object so that the actor understands what he is able to achieve with it. From this assimilation, he comes to appropriate it, to build a set of knowledge and skills associated with its functioning. The loop is completed by the actor's reflexivity towards the artifact, which then opens up new opportunities relating to its use or in relation to the actor's needs. It is a dynamic process that is mainly based on action situations in which the object is used and that feeds the understanding of the user/actor.*

From these constructions emerge the situated actions that are associated with the generation of organizational knowledge then assimilated to "an organizational heritage of symbolic, communicable and storable representations, mental models of actors transferred on material formats" (Lorino 2005, p. 61).

This question of devices is part of the context of organizations where a collective intelligence can then express itself considered as "a utopia, a project, which aims at the emergence, thanks to the digital revolution, of an intelligent collective, a real time democracy (direct democracy assisted by computer or virtual agora), a blossoming of people in the exchange of knowledge, a decentralized, transversal and not hierarchical" (Berthoud *et al.* 2002, p. 33) where the human being tries to achieve ubiquity. Collective intelligence then becomes an exploitative lead on the question of the transformation of the organization, both on the social question and on that posed by technology.

3.2. Collective intelligence, an operating model for organizational transformation

Collective intelligence is marked, in its current acceptance, by the technical tools that contribute considerably to its emergence. However, this notion is ancient and has evolved to its different current meanings such as "wisdom of the crowd"[3], the Noosphere[4] and the pragmatic approach of collective intelligence, even stigma[5]. In the case of this book, we will use the following definition of collective intelligence: "an intelligence that is distributed everywhere [...], an intelligence that is constantly valued [...], the real-time coordination of intelligences [...], leading to an effective mobilization of competences" (Lévy 1997, pp. 29–30). The question related

3 This current was built in opposition to the negative view of the crowd developed by Le Bon (1905). Authors such as Lévy (1997, 2002, 2003), by considering that "collective intelligence has nothing to do with the foolishness of crowds" (Lévy 1997, p. 87), offer another reading that is supported by the development of the notion of wisdom of the crowd (Surowiecki 2004; Arazy *et al.* 2006), as a specific branch of study of collective intelligence.

4 Concept described by Teilhard de Chardin (1995) in which he evokes a particular conception of a "collective planetary consciousness" that is the Noosphere. If this concept is deeply marked by religion in Chardin, De Rosnay (1986, 1995) reuses the Noosphere in a neo-Darwinian and biocybernetic approach, as a "planetary brain, a 'collective intelligence', an informational ecosystem resulting from the multimedia (or multimodal?) interconnection of all dematerialized human productions and in exponential growth due to the convergence of the computer revolution, biotechnologies and neurosciences" (Jouët 2011, p. 40).

5 This biological approach is based on the collective behavior of ants considered as an example of collective intelligence.

to collective intelligence is whether, together or alone, human beings are more "intelligent" or to use Lévy's question: "can human groups be collectively more intelligent, wiser, more imaginative than the people who constitute them?" (Lévy 1997, p. 240). From a pragmatic perspective, several characteristics are identified (Maleszka and Nguyen 2015):

– autonomy of the members, in their choices, decisions, etc. It should be noted that in some current discourses (Roux 2010), the autonomy of individuals is reinforced by the use of ICTs, which would be a factor of emancipation, or even "empowerment"[6] (Bernard 2014). Current organizational trends around self-organization, especially of employees (Lee and Edmondson 2017), are moving away from the traditional model inspired by the military or Taylor (Autissier *et al.* 2016) and towards the network. According to Lee and Edmondson (2017), three facts motivate this search for less hierarchical organizational structures: the acceleration of information flows and technological innovation in an uncertain context where the rigidity of the hierarchical company does not offer an adequate response to market stimuli (new entrant, new regulations, etc.), and the need for a more flexible organizational structure; the development of work based on knowledge and the associated knowledge economy (the intangible, such as expertise or creativity, is then the main source of wealth creation); and finally, the growing demand for meaning from employees in relation to their activities;

– presence of complementary knowledge, so as to encourage a plurality of points of view and thus effectively implemented responses;

– orientation towards a common goal. This last feature is a "symbolic cement" (Cordelier 2016). Indeed, the organization manages to maintain its organizational coherence thanks to the common pursuit of objectives, which then justify the actions put in place by its members.

The organization presents itself as a fertile ground for the study and analysis of behaviors based on collective intelligence because of its ability to generate collective tasks and goals. This has reached the point where today collective intelligence is accepted as an asset for the organization by understanding it as a decentralized approach that allows for community-based problem-solving.

---

6 This term emphasizes the empowerment of the individual, and some people also translate it as "autonomization". It has already been discussed in Chapter 2.

3.2.1. *Origin of the concept of collective intelligence*

Lévy (1997), in his book *Collective Intelligence*, identifies among the first research works on collective intelligence those of a few Arab theosophists who lived between the 10th and 11th Centuries: Al-Fârâbî (872–950), Ibn Sina (better known as Avicenna) (980–1037), Abû'l-Barakât al-Baghdâdî (1080–1164) and Maimonides (1135–1204). Their conception, presented and analyzed by Lévy, is based on the idea of a single intelligence (the same intelligence would be shared between all human beings), but separated between different individuals; it is this intelligence that would link humanity to a transcendental being.

These authors provide a framework for Lévy to think about his contemporary conception of collective intelligence as intelligence distributed to the whole of humanity. In his work of analysis and instantiation of collective intelligence, he makes a transposition so that "this which was theological becomes technological" (Lévy 1997, p. 96), that is, humans would be connected and that they would reach, through technology, a transcendental level. The space in which this intelligence is placed is then composed of virtual worlds that "propose as instruments of self-knowledge and self-definition of human groups, which can then constitute themselves as autonomous and autopoietic collective intellectuals"[7] (Lévy 1997, p. 101) (author's translation).

If the first approaches of collective intelligence were able to develop without technological input, today, the place of networked computing, especially the Internet, is preponderant. Work on collective intelligence benefits from research related to computer science, such as that on collaborative tools or "wikis"[8]. Through this networking of intelligences[9], it becomes possible to coordinate numerous participants. With ICTs, it is a collective intelligence increased by digital technology that deploys it; these

---

7 This is a concept developed by Maturana and Varela, who define an autopoietic system as a "network of processes producing components that (a) continuously regenerate through their transformations and interactions the network that produced them, and (b) constitute the system as a concrete unit in the space where it exists, specifying the topological domain where it is realized as a network" (Varela 1989, p. 44).

8 The wiki is a computer concept popularized by Cunningham that represents a collaborative website where everyone is able to contribute to the content of the site. The Hawaiian term from which it is taken (*wikiwiki*) means "fast".

9 De Kerckhove (2000) entitled his book *L'intelligence des réseaux*.

technologies carry out an action of "distribution" of the minds of the various members in a global network. This is more or less what Douglas Engelbart[10] wished to implement, who conceived the computer as "a technology that allows the minds of individuals to be distributed in the technical system with which they co-evolve: computer calculation frees people's creative energy by revealing to them the common ground that binds them together beyond the ordinary forms of interaction" (remarks reported by Turner (2012, pp. 19–20)).

3.2.2. *Instrumented collective intelligence*

The authors who focus on organizations as a place of expression and construction of collective intelligence are questioning how the latter can be set up, the tools that participate in its constitution or even its contribution to the organization. Most of the works are thought in a pragmatic logic, especially with computer science, even though the virtual is neither the source nor the goal of this collective intelligence. The counterpart of this approach is research that is part of the continuity of the Noosphere (Chardin 1995) such as the *Global Brain* (Glenn 2013; Heylighen and Lenartowicz 2017) which is defined as a self-organized and adaptive network formed by all the inhabitants of the planet and the ICTs that link them in a global system. The underlying idea is that individuals are, in a globalized context, so interdependent that the whole can be understood as a superorganism[11] composed of both organs – in this case society as a whole – and components that are individuals.

Favored by the convergence of information technology and tele communications, technical tools thus contribute to the emergence of an instrumented collective intelligence (Laniau 2009; Lykourentzou *et al.* 2011; Comtet 2012; Skaržauskienė *et al.* 2014). Turner (2012) highlights this community dimension in his book on the history of the Internet and the first physical communities that became virtual communities. Thus, with the *Whole Earth Catalog*[12], Brand's desire is to bring together different communities

10 Researcher at the Stanford Research Institute, he is a major contributor to the development of computer science with his work on the mouse (of which he is the creator) on hypertext and also on collaborative tools.

11 This then echoes the considerations of Chapter 2 on the platform and its association with the concept of superorganization.

12 Stewart Brand has made a major contribution to the history of IT with the *Whole Earth Catalog*, with his collaborations with Engelbart as well as with many structures he created such

(especially from the counter-culture) in a single space. The meaning of this commitment continues with the digital transition of the *Whole Earth Catalog* becoming the *Whole Earth 'Lectronic Link* (WELL). It is in this context that the term "virtual community" is used for the first time by Rheingold in an article in the *Whole Earth Review* (linked to WELL). Indeed, we are in the case of "human–human" interactions that rely on the technical tool as a mediator. This ideal has influenced the orientations of digital technologies and *a fortiori* in Silicon Valley where WELL members are very present. Thus, researchers at the Stanford Research Institute in the 1960s (in particular Engelbart, linked to Brand) wanted to create a system where "everyone that could understand would be enhanced by the participation of others through a collective *feedback* process made possible by computer" (Turner 2012, p. 65).

The link between ICT and human intelligence is particularly strong among contemporary thinkers of collective intelligence; indeed, these technical tools "have widened the potentialities since it is now possible to work together, on a very large scale, on the same project, but at a distance, synchronously or asynchronously, on modifiable media, while allowing a direct relationship between 'collaborators' or 'cooperators' thanks to the Internet" (Gangloff-Ziegler 2009, p. 95) and they possess "capacities to foster these processes [issues of individual and collective time and space] inherent in any organization" (Commandré 2011, p. 266). In order to study the way ICTs and humans interact, a line of research has been built, that of human–computer interaction (HCI) (Ng 2002; Bigham *et al.* 2014), which can be related to cognitive ergonomics and work in ergonomic psychology. The authors in this field then identify three main lines of study of collective intelligence (which they reduce to the "wisdom of the crowd" by focusing on crowdsourcing, which is literally supplied by the crowd) which are combined with human–machine interactions (Bigham *et al.* 2014, p. 2):

– "directed crowdsourcing, where a single individual attempts to recruit and guide a large set of people to help accomplish a goal";

– "collaborative crowdsourcing, where a group gathers based on shared interest and pursue that goal together";

– "passive crowdsourcing, where the crowd or collective may never meet or coordinate, but where it is still possible to mine their collective behavioral patterns for information".

---

as the Global Business Network or to which he contributes such as The Long Now Foundation (funded in particular by Jeff Bezos, founder of Amazon).

However, this approach appears limited in that it restricts it to the logic of crowds and a micro-division of tasks among the members of this digital crowd. However, this work on human–machine interactions is essential in the sense that "in the collaborative environment, full physical co-presence is lacking. Thus, the logic of use of certain ICTs cannot avoid implementing a real work of transmission and understanding of interpersonal information. Individual and collective cognitive performance is at stake" (Comtet 2010, p. 131). This also makes it possible to shed light on the processes underlying human–machine interaction through the mediation of ICTs and in particular the misuse of them.

Currently, in connection with the exponential growth of digital data and information produced (Chen and Zhang 2014), collective work is all the more necessary to deal with this available mass. Thus, Jung (2017), in his editorial, combines collective intelligence and massive data processing for the collection, analysis and use of this data. However, this subject was addressed at an early stage by Lévy, who put this issue into perspective by proposing as a solution to promote "the socialization of problem solving rather than its separate processing" (Lévy 1997, pp. 68–69).

3.3. Collective action and analysis of the potential of collective intelligence

In order to study the practices and uses inherent to a collective intelligence, the context of emergence cannot be ignored, and in particular, the methods hitherto used in the organization to encourage or, at the very least, support collective action. Thinking about collective intelligence runs parallel to current technical discourse, which suggests replacing the human with algorithms (classification, neural network, Gauss curve, etc.). For his part, Lévy states that "the best thing that can be done with new technologies is not artificial intelligence (AI), but, on the contrary, collective intelligence: that computers do not imitate humans, but help them think and evolve their ideas collectively" (Lévy 1997, p. 36).

3.3.1. *Setting up collective action*

The communication system in the organization is presented as a formalized system of collective actions that also reveals the complexity of coordinating these actions at a distance (Alag 2009; Castro Goncalves *et al.*

2011). How do the members of the organization coordinate themselves in order to build collective action in the organizational schemes of their entity? How is this collective action situated, dependent on the context, nature and result of interactions, and constructed through the technological mediation of ICT? By mobilizing collective action and intelligence to build the organization, it is possible to concretely implement an organizational change, so as to evolve from a positivist and hierarchical conception of the organization to a decentralized and horizontal organization.

Collective action is built and structured within the organization, where organizational action takes precedence (Teulier 2005), with the aim of solving a problem or achieving an objective. Crozier and Friedberg (1977) emphasized the importance of collective action as a social construct in our society. Indeed, for them, it is a human artifact that contributes to the orientation of behavior and thus to the capacity for action and freedom of individuals. This action then represents "an association (obliged or decided) of actors in order to cooperate, to face a problem and in a general way, to uncertainties" (Comtet 2007, p. 99). Collective action is thus a concrete materialization of collective intelligence expressed in a group.

Comtet also stated that "this level of analysis is essential insofar as the organization expresses or not the willingness of those in charge to set up mediated, synchronous or asynchronous communication systems, and to set up rules specific to a work that is intended to be more or less common between all the players present" (Comtet 2007, p. 99). Thus, to carry out this specific action, the protagonists mobilize a collective competence which is "defined as the result of an interactive process between formalized knowledge on the one hand and tacit knowledge on the other, [and] becomes an evolving system of knowledge shared and applied by the members of organization" (Batazzi-Alexis 2002, p. 6). The link between collective action and organization is such that it is possible to consider that they are "the two inseparable faces of the same problem: that of structuring the fields within which action, any action, is developed" (Crozier and Friedberg 1977, p. 22). Moreover, collective action can be understood as a system of power through the creation of these spaces of freedom for the members of the collective and the answers they will bring to the problems raised, which are then shared knowledge.

3.3.2. *Collective action and organizational knowledge*

The collective action, which is "situated" action (Suchman 1985; Guyot 2009) where the behavior of actors is linked to that of other protagonists, "lies at the heart of the problem of knowledge in organizations; it raises the question of activity (in its individual and socially organized dimension), of the meaning imparted to activity by actors and of the activity engaged in by actors as a result of their judgements: of meaningful action and effective meaning" (Lorino 2005, p. 56). From individual knowledge, knowledge can move to a community or even organizational stage in this collective framework (Hassel 2007) as illustrated by the SECI model (Nonaka and Takeuchi 1995) (see section 2.4.2). The last two stages of the model are a sharing of knowledge that is combined with the other knowledge of individuals or a community of practice, before it is internalized by the individuals who need it. Collective work thus promotes the sharing of experiences and knowledge that eventually becomes tacit knowledge (Baujard 2011a). The idea is found in a conception of inseparability between collective action and knowledge (Hatchuel 2005, p. 80) where both are mutually constructed in the organization. Indeed, collective action, which is situated, is context-dependent as a result of interactions; there can be no generic response. The aim here is to develop an understanding of the transformation of collective action thanks to the ICTs mobilized in its realization and coordination, as illustrated by Dudezert *et al.* (2014) with the introduction in an organization of an information system dedicated to knowledge management. This new system thus leads to an evolution of collective action and its associated meaning. In the case presented, the authors underline the necessary evolution of organizational knowledge and its management during the introduction of this new technical system.

The knowledge from the group contributes to the development of this collective and is used by all those who need this knowledge. Knowledge then becomes an intangible common good that everyone can contribute to and to which everyone has access; the use of a knowledge management system to store and manage the group's intangible assets is of particular interest when employees arrive or leave (Biron and Hanuka 2015). The knowledge resulting from this approach is then a resource available to the members of the community concerned. New informational routines or modes of cooperation can then be set up (El Amrani 2008), individually or collectively, such as daily consultation of the organization's digital social network.

Relational networks such as communities are thus identified as favorable factors for the emergence of coordinated actions that are the beginnings of a collective intelligence.

3.3.3. *Framework for analyzing the potential of collective intelligence*

The need to better understand (or even succeed in materializing) collective intelligence can be satisfied by the framework for analyzing the potential of collective intelligence developed by Mačiulienė *et al.* (2015) (Figure 3.2); the authors detail the components of the different dimensions that allow them to move towards collective intelligence[13].

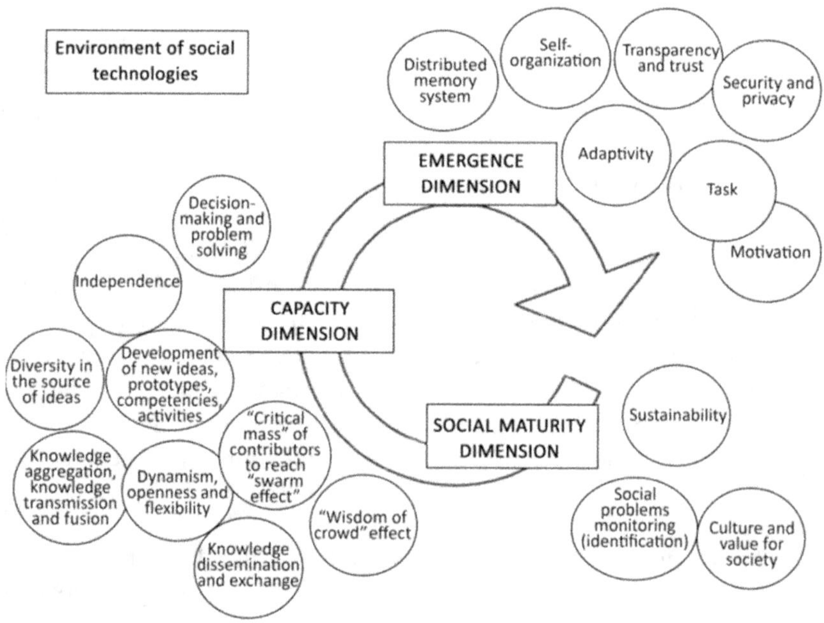

Figure 3.2. *Framework for analyzing the potential of collective intelligence (according to Mačiulienė et al. (2015, p. 1719))*

13 We have concretely applied this analytical framework on a community of developers. The article is available at the following address: https://www.openscience.fr/L-emergence-d-une-Intelligence-Collective-dans-une-grande-entreprise-de-l.

COMMENT ON FIGURE 3.2.– *This analytical framework is the result of the work of the three researchers with a view to highlighting (or even successfully predicting) the emergence of a collective intelligence. They identify three main dimensions that are broken down into factors and questions (details of the questions are available in the Appendix) that are used to analyze a collective in a fairly exhaustive manner:*

– capacity: *that is, the ability of the community studied to generate action and knowledge located in relation to it. The diversity of profiles and the openness of the community must allow for the original production of knowledge to meet the needs of members;*

– emergence: *it is based on the capacity of the collective to self-organize. Self-organization is based on effective equity and equality in the structure. Emergence can also be linked to the setting up of a dedicated information system to share knowledge and is reinforced by the trust that may exist between the contributors;*

– social maturity: *for the authors, this dimension comprises an essential part of the community's collective intelligence potential because it ensures the survival over time of the collective intelligence and helps to overcome the problems encountered (mainly related to the human factor such as conflicts between members and changes in the number of members that influence the structuring and orientation of the community).*

The application of this analytical framework helps to better understand the collective intelligence within an organization; the issues and dimensions identified by the authors help to better circumscribe communities and their potential, especially in organizations that may be highly decentralized.

3.4. From communities to meta-organization

Decentralization, driven by new technologies and promoted by collective intelligence (Huang *et al.* 2017), is an operation that is "fruit of initiatives taken as close as possible to those who produce and use information, raises risks of inconsistency or failure in terms of maintenance" (Guyot 2000, p. 99). The hybrid model presented by the author, "centralized in its structure and through common codified procedures, ensures general consistency, while decentralizing production, with producers' initiative being framed by technical or common editorial procedures" (Guyot 2000, p. 99); it seems more relevant in an organization by ensuring an overall vision. This collective intelligence in the organization supports the decentralizing effect of ICTs and accompanies the transition to a networked organization.

This model can help to strengthen the link between cultural and social approaches to knowledge organization, through the influence of culture (López-Huertas 2016), providing the basis for the design of knowledge organization instruments in line with the different strata of the organization. In fact, this approach promotes an awareness of the diversity and/or complementarity of forms of action (or even opposition) and thus of the knowledge needs of the protagonists.

Collective intelligence is then consubstantial to the organization and at the same time participates in its construction. This inscription of collective intelligence in the organization is particularly expressed in the form of communities of practice which "are par excellence the space of procedural rationality and transmission. The members come to consolidate practices and their identities are inseparable from the mastery of these practices" (Zacklad 2016). It is through these communities that the members of the organization build meaning together.

3.4.1. *Communities of practice*

Communities of practice can be a breeding ground for the emergence of collective intelligence in organizations. Indeed, activity within organizations is based on a collective basis (in connection with the division of labor) that is composed of routines, projects and actions that make (or do not make) sense.

Theorized by Wenger (1998) and taken up by many researchers (Kimble and Hildreth 2005; Dameron and Josserand 2007), communities of practice are described as "groups driven by mutual obligations and a shared passion for a particular subject" (Tessier *et al.* 2014, p. 123). These are what we can call intra-organizational networks that play two main roles:

– exchange and production of new knowledge and common practices; thus an organizational and knowledge management role (Teigland and Wasko 2003). Regarded as the "shadow system of learning" (Stacey 1996, p. 1) in organizations, communities of practice are joined by people who want to learn something specific and want to contribute to knowledge production communes;

– creating and strengthening the social bond between members of the organization through member engagement and investment in the community of practice (Wenger *et al.* 2002). This also helps to develop a sense of belonging in order to limit employee departures.

Through the work on communities of practice, three main characteristics stand out (Wenger 1998):

– their area of knowledge: this is the factor that encourages members to come together to form a community of practice. They define their knowledge object, the aspects they want to work on and the problems they want to solve. They invest themselves according to their interest in this object of knowledge;

– how they function: by creating the community of practice, members specify the relationships they maintain, the means they use, the frequency of work and promote trust between them. Communities of practice can be more or less formal, especially when they are less related to the main profession of members;

– their capacity to produce the practice, also called "shared repertoire" by Wenger, i.e. the resources they have at their disposal, their routines, their productions, as well as elements of language and the rules of the community of practice. Indeed, a community of practice is above all a place of socialization and collective production.

In their research, Wenger *et al.* (2002) also identify five steps that lead to the formation of a community of practice:

1) potentiality: this dimension is reflected in the willingness of actors who collaborate informally and who gradually structure themselves in the form of a community of practice by sharing their problems, interests or knowledge. Factors such as passion or shared values are as important as the

assumption of responsibility by some members to help the development of the community of practice. Three dimensions are essential: identifying the topic addressed by the community, identifying potential community members and identifying knowledge needs;

2) grouping: once the first step has been completed, it is possible to formalize the existence of the community. It is then necessary to set up activities, exchange times, etc. in order to create trust between members and to highlight common needs/interests;

3) maturation: the community can grow consistently and rapidly. It is essential (still for Wenger and his co-authors) to support its growth to ensure its sustainability and to succeed in focusing on its raison d'être;

4) organization: the organization of the community is key in order to reduce the natural tendency to divergence and dispersion. Likewise, it is necessary to maintain momentum to ensure that the community remains active;

5) transformation: naturally, the community, as a living organism, should tend to evolve (in connection with new ideas, new members, the establishment of another orientation, etc.), or even disappear if the community no longer has a reason to exist.

Communities of practice, as human collectives and dependent on the environment in which they operate, are increasingly mediated by ICTs to the extent that the so-called "virtual" communities of practice are emerging, which can also be virtual communities of practice.

3.4.2. *Virtual communities*

In the case where community members are deterritorialized and use ICTs to organize and exchange, we then speak of "virtual communities" (Dubé *et al.* 2005; Casteignau and Gonon 2006). This "specific form of intra-organizational network allowing the exchange and development of knowledge and practices" (Tessier *et al.* 2014, p. 121) was popularized by Rheingold (1993). It is now recognized as a space for knowledge production and management for organizations (Chou and Hsu 2018).

The specificity of the virtual community stems from the fact that "it operates via remote communication technologies, [which] are commonly used as a knowledge management tool, in many multi-nationals, thus becoming a standard rather than an exception" (Tessier *et al.* 2014, p. 123). It therefore uses the tools made available by the organization, not hesitating to modify/divert the primary use of the tool. This modifying action illustrates the existing dichotomy between the basic presupposition (what is supposed to be done with the device) and what is done in practice (the catachresis phenomenon mentioned in section 1.4.3).

This raises the question: is knowledge more easily accessible in a virtual community than in a community of practice? Shin and Kook (2014) developed an approach where they focus on knowledge through what they call Virtual Knowledge Organizations Network (VKON), which they define as a working group of knowledge workers sharing topics/interests/problems/common experiences/practices and knowledge. The importance of this type of community is great for organizations by fostering, for the authors, a link between knowledge management, online collaboration and bringing together knowledge actors in the organization, which can result in the establishment of virtual communities of practice.

3.4.3. *Towards virtual communities of practice*

This relationship between real and virtual then makes it possible to link communities of practice and virtual communities to lead to virtual communities of practice (Tessier *et al.* 2014). Indeed, collective practices are increasingly mediated by sociotechnical devices, which are used to meet the expectations of communities and the organization in relation to them. It is possible to accept virtual communities of practice as a "sociotechnical system" providing the energy needed for effective and technically efficient social development (de Moor and Weigand 2007). Indeed, virtual communities of practice are, in a professional setting, very specific structures where members participate in a common understanding of their field of action and carry out tasks related to their community through ICT (Bieber *et al.* 2002).

The moments of meeting and exchange are essential to maintain trust between members and towards the organization. Indeed, unlike offline communities, trust in communities where exchanges are mediated by digital devices relies primarily on written communication (Ridings *et al.* 2002). Indeed, in some communities, members may never meet face-to-face. Trust differs in these groups where members may never meet, being not only inter-individual trust but also collective trust, with shared knowledge accessible to all. Two kinds of trust are then identified in practice:

– *community* confidence, relating to the construction of a sense of belonging: this is concretely materialized by the implementation of unifying symbols (logo, manifesto, specific events, etc.);

– *institutional* trust, relating to the link between the community and the organization in which it is embedded.

These two forms are complementary in addressing both the link between community members and the link between the organization and the community. This notion of trust is fundamental: it is the "key" to encourage members to get involved (Ridings *et al.* 2002). A climate of trust then facilitates the sharing of knowledge between the different protagonists and thus enables the community to play its full role in managing knowledge.

These autonomous communities are part of a more horizontal[14] logic (Bernard 2015) from the moment they think they are linked to the achievement of the organization's goal. They are built on a logic of autonomy and network. This need and willingness to share cognitive resources is such as to avoid the wanderings that consist in the creation of "systems allowing to meet the specific needs of this department, taking very little or no account of the information necessary for other departments. [...] Often, department heads were attached to this kind of system because they could control the data: it was their data that were not considered as collective resources, which should be easily shared" (Botta-Genoulaz 2007, p. 40). Contrary to this logic, sharing is fundamental to the success of the approach. In this way, knowledge production and its management are located in the virtual organization. This can be part of a logic of intra-organizational sharing via networked computing.

---

14 As opposed to a vertical, highly hierarchical logic. A horizontal organization is supposed to promote decentralized decision-making, stakeholder autonomy and accountability.

3.4.4. *The construction of a meta-organization through communities*

It is the result of the sociotechnical devices, with their "logic from use" (Perriault 2002) – used by communities of practice and virtual communities of practice in line with the goals and constraints of the organization, as well as the involvement of the members – which leads to the construction of a meta-organization as collective intelligence. If we follow the logic of Rabardel (1995a, 1995b, 2005) on the construction of an instrument, then we are in the phase of instrumental genesis of a meta-organization.

We define meta-organization as *a specific form of reticular organization that is built from the communities of practice and virtual communities of practice that compose it, which it links and directs to achieve its goals.* We can compare the meta-organization to an organism that would be composed of multiple sub-organizations cooperating and acting symbiotically so that the overall organism can act (Figure 3.3).

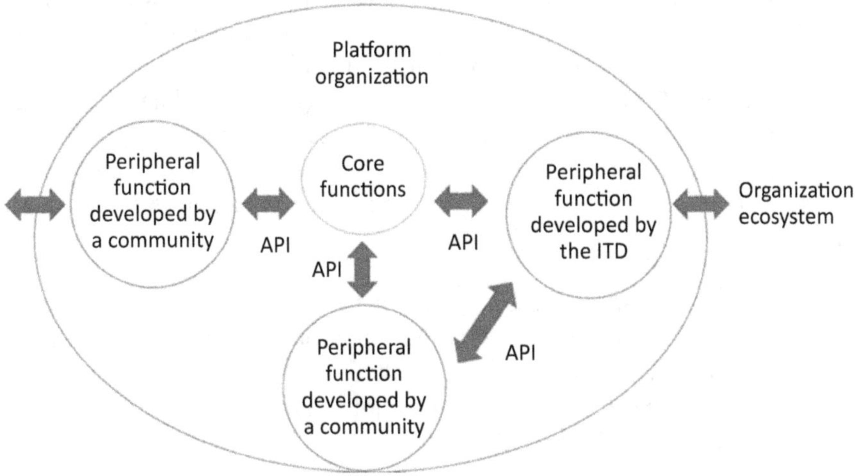

Figure 3.3. *Representation of the platform organization*

COMMENT ON FIGURE 3.3.– *In this schematic representation of the functioning of the meta-organization, virtual communities of practice and communities of practice contribute to the various peripheral functions that make up the meta-organization and to the achievement of overall objectives. Within the framework of instrumented relationships, the exchange can be done through application programming interfaces (or APIs) that facilitate the interoperability of information systems between them. ICTs play a key role at the level of action and collective intelligence where members can move from one group to another and access the knowledge needed for action. This opens up new fields of action and relationships that reconfigure the organization through physical or virtual communities engaged in a logic of exchange and collaboration.*

In the case of this meta-organization, several aspects are to be taken into account in connection with the inherent characteristics of collective intelligence:

– the hierarchical structure of the organization: depending on its degree of horizon or verticality, the levers that can be mobilized and the freedom left to the different actors will not be the same;

– the setting up of the famous "symbolic cement" (Cordelier 2016) on which depends the ability to federate the actors around common objectives and which make sense: the context in which the members of the organizations evolve associates at the same time infrastructures (premises, data processing, etc.), communication and collaboration protocols (use of digital social networks of companies or even tools qualified as collaborative) coupled with processes (decision-making in particular) and common resources (knowledge, budgets, etc.).

These criteria provide a basis for the creation of an environment of trust and mutual support necessary for the meta-organization to fulfill its missions. The whole is also supported by these technical infrastructures that are a technological base supporting interactions, as well as providing access to knowledge with knowledge organization and management systems.

With this sociotechnical instrumentation of collective intelligence, the model that emerges is that of meta-organization as a decentralized organization that allows for community-based problem solving and brings together several intelligences within a single entity. Thus, the members of the organization are producers of knowledge internally and for their peers.

The concrete conservation and exploitation of this heritage, resulting from the situated actions to which they respond, is an essential issue for these communities. Through their actions and their commitment to the organization, community members play a role in the organization's transformation, because from being individuals, the experience becomes collective, each one bringing his or her contribution to the organization.

4

Organization Experience

The different notions mobilized in the previous chapters (such as sociotechnical devices, collective intelligence or platform/meta-organizational structuring with an anthropocentric orientation) converge towards what we call "organization experience" (OX). This is the worst of "user experience" (UX) (Chevalier and Kicka 2006; Hassenzahl 2008), which has emerged in connection with the unprecedented developments that each user specifically maintains with the information and communication devices – commercial or not – that accompany his/her daily life, in the digital and information age. At the individual level (with the UX), the objective is to carry out a reflection at the organizational level with the OX. Our path relating to the OX first returns to the question of organization and change, before highlighting the capacity of sociotechnical devices to bring about organizational transformation. It will be completed by the social layer through the effective mobilization of collective intelligence. Finally, we will end by the use within this organizational form of digital common goods, and more particularly the question of the interoperability of the systems underlying the whole.

4.1. Organization through the prism of change

The organization is subject to constant transformations in the course of the informational and communicational processes that nourish it and that it contributes to producing[1]. These depend on the influence of personnel,

---

1 Changes that may take the form of the concepts presented beforehand, although it is not possible to limit organizational changes to these concepts alone. They may also result from

interference from stakeholders[2], and structural or technological pressure. It is then a way of questioning three dimensions of the organization:

– the social dimension: in particular the relations maintained by the individuals working together;

– the functional dimension: the functioning of the organization, its rules, its structure, etc.;

– the epistemic dimension: linked to the representation of the organization produced by the activity of its members.

These three dimensions are subject to permanent tensions, as the organization operates by nature in an unstable environment. Thus, in the social dimension, staff involvement is focused on activities that are expressed in risks (financial, strategic choices, transfers/purchases, etc.) taken in the context of these changes.

In the case of actors with an extremely rich industrial, social, cultural and technical heritage, changes are likely to lead to critical situations. Indeed, by placing itself in a zone of discomfort (e.g. the evolution of the structure, the introduction of new devices or the cessation of certain activities linked to the automation of a task), the organization is confronted with inertia (Table 4.1 illustrates five types of inertia), which then makes its evolution more complex.

Internal organizational changes (which address both the functional and epistemic dimensions) come from the directorates and departments. With the massive use of ICTs, the source of these changes often lies within the information technology department (ITD), which initiates, or *at least* supports, these changes. Thus, a current organization is characterized by a continuous construction or commissioning of[3] sociotechnical devices, as analyzed by Guillemot and Kocoglu (2010). These authors identify, in particular, a logic of accumulation of systems, a recurrent situation where IT departments constantly deploy new devices with a view to responding to the

---

exogenous factors (such as regulatory constraints) or endogenous factors (such as the company's business strategy or salary demands/unions).

2 Under this name, we cover all the actors who have a relative interest in the organization in question (Bonnafous-Boucher and Dahl Rendtorff 2014).

3 In cases where the organization relies on an IT service provider, a situation that has been reinforced with the implementation of Software as a Service (SaaS). In this configuration, the software service is accessible from a web browser and no longer directly on the user's computer.

demands of their users (even if it means generating cognitive overload through the multiplicity of information systems) or automating tasks.

Types of inertia	Central concept	Agent design
Psychological	Denial and fear of learning	Agents are overwhelmed by their negative emotions in response to the perception of a threat
Socio-cognitive	Standards and values at individual, group, organizational and societal levels	Agents are integrated into institutions characterized by their adherence to norms and values
Sociotechnical	Technological and sociotechnical dependencies	Agents are integrated into sociotechnical systems that have their own dynamics, especially on the grounds of time to the development of these systems and their internal coherence
Economic	Economic dependence	Agents are integrated into business models that have their own dynamics resulting from redistribution of resources between the exploitation and exploration processes
Political	Interest earned and alliances	Agents are integrated into networks of vested interests, which have their own dynamics, especially in view of the time required for the alliance building

Table 4.1. *The five typologies of inertia existing in organizations (according to Besson and Rowe (2011))*

COMMENT ON TABLE 4.1.– *This table summarizes the literature on organizational inertia according to Besson and Rowe (2011). They identify five main categories of inertia related to the (non-)transformation of an organization. These are then supposed to cover the various obstacles to change and thus facilitate the construction of an adequate response to the situation.*

4.2. Sociotechnical device and catalyst for organizational transformations

Technical mediation and the increasing digitization of activities (and reality) through such devices are thus becoming the norm. The growing influence of this fact leads us to question the roles, the underlying logic and the expectations of these systems. Indeed, by their coercive nature[4], they contribute to the transformation of the organization. Hence the growing importance of the question: which actors are building devices in an organization? Thus, in large network organizations, three situations can coexist:

1) Internal development within the information technology department: the ITD is responsible for producing the systems required for the organization to function properly. To this end, it is equipped with internal teams capable of carrying out this work. This situation is frequent in organizations whose core business is linked to digital technology.

2) Development by an external organization: delegated to a computer engineering services company, also called a digital services company, development focuses on technicality (algorithmetic or methodological) so that these service providers "generally evade questions related to the conditions of use and mismanage – when they do not ignore them – the ergonomic, semiotic, and more generally usability dimensions of the realized devices" (Papy and Sansonetti 2014, p. 33). This orientation exacerbates SaaS and platformization, since the organization contracts with the external service provider who provides development and proposes *ad hoc* technical infrastructures.

3) Development internal to the organization but outside the ITD: in this configuration, the person in charge of development may lack technical skills for advanced functionalities or ergonomics. However, his or her knowledge of internal needs and practices helps to compensate for these shortcomings through a detailed understanding of the ins and outs of the project through a symbolic system of representation (Bélisle 2002) shared with end-users. Not recognized by the ITD governing bodies, as they are assimilated to "shadow IT" (Silic and Back 2014; Silic *et al.* 2016), the computer programs they develop are, however, used on a daily basis in organizations[5]. In this case, three functions are identified for the devices produced (Agostinelli 2009). The first is organizational: these tools apply to situated knowledge; it is correlated with the context of its use, it is "glocal" knowledge where "it is actually about recreating, for each level, a vision that is consistent with the global vision [...]

4 Indeed, once transposed into a device, a practice is "frozen" through its transcription into computer code, on which users of computer applications rarely, if ever, have the possibility to intervene (Flichy 2003, 2008). Modifications can only occur during future updates of the system or by hijacking the device.

5 Thus, in the article by Henry (2019b), we return to a community of developers who are building information systems outside the framework of the ITD and which are nevertheless re-embedded in the company's global program.

This means making the local consistent with the global ('the locbal') and bringing up to the global level what emerges from the local ('the glocal')" (Lenhardt and Bernard 2005, p. 82). The second is informational: tools are the source of mediation between knowledge and practices. The third is regulatory, since it is a question of synchronizing the functioning of the first two. To these, Agostinelli adds a function of strengthening the sense of belonging to the community through the development and use of specific tools that meet their needs.

To deepen our understanding of the choices made, it is also necessary to put the context of ITD action into perspective (Figure 4.1).

In this book, we will insist on the logic of construction in use (Folcher 2015), a situation that is mainly expressed in situations 1 (in-house development) and 3 (in-house development outside the ITD).

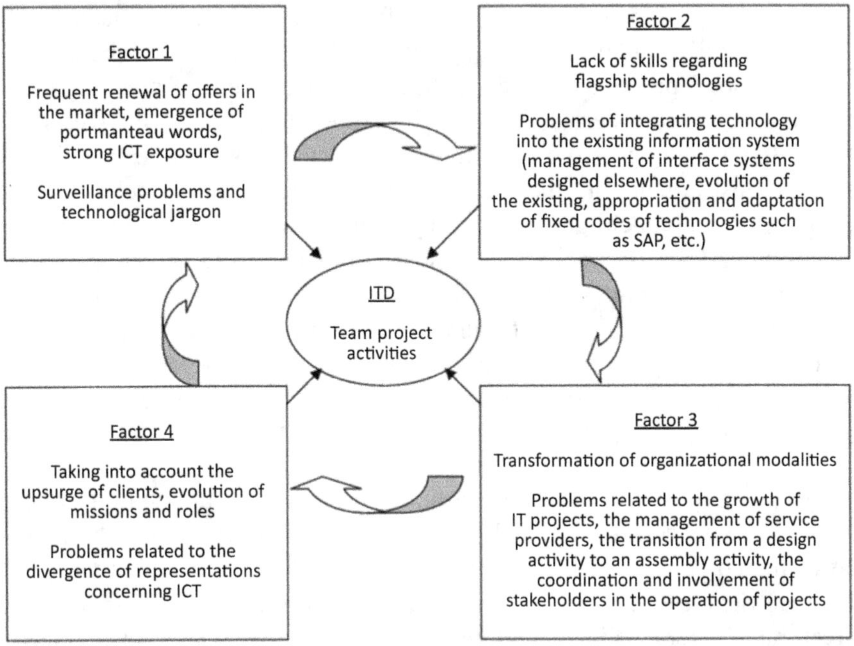

Figure 4.1. *The context of ITD action (according to Castro Goncalves (2011, p. 41))*

COMMENT ON FIGURE 4.1.– *This representation of the context in which current ITDs operate helps us understand why they increasingly use providers (factors 1, 2 and 3). Indeed, the speed of technological change is such that ITDs are quickly overwhelmed and rely on service providers who offer (more or less) trained staff to meet their operational needs. Factor 4 also favors IT development by operational staff to meet very specific needs. This is all the more true since the ITDs cannot address all the technical needs of the members of the organization; this state of affairs is at the root of the creation of the community of local developers, who then respond to unmet needs. Thus, the question that emerges is how to collaborate within the same organization so that the different actors can count on the skills and means in adequacy with their needs to meet the missions that have been set for them, while taking into account the legal and patrimonial issues of the organization (protection of tangible or intangible heritage).*

4.2.1. *Co-construction of sociotechnical devices*

The question of the device's relevance for end-users arises and leads to the effective mobilization of the teleological principle to ensure that the device's aims, the users' problems and the collective construction of meaning are in line with each other (Verlaet 2015). Indeed, the device retains the sole status of an artifact as long as the conditions are not met for it to be transformed into an instrument at the service of anthropocentric activity. The technocentric approach, guided more by "what it is possible to do only from what it would be desirable to consider" (Folcher 2015, p. 40), and the hierarchical injunctions that come to "force" the artificial transformation of the artifact into an instrument cannot ignore the cognitive imperatives of appropriation inseparable from the instrument: "personal scheme of use or [...] dominant social pattern of use that the person has transformed into a personnel pattern" (Bélisle 2002, p. 28).

Following the choice of the actor to develop a technical solution, we then address its actual construction and underlying logic. One evolution observed is that of the co-construction of the sociotechnical devices, a logic that is part of a historical evolution that mirrors the transformations in France of the Directorate General of Telecommunications (which became France Telecom and then Orange), according to Hochereau (2006). The author highlights three successive orientations for the construction of the systems[6]:

6 The author uses the "piloting computerization" syntagm.

– A mode thought in an organizational logic, supported by business informatics. It reflects a desire to rationalize activities and optimize their coordination[7]. This rationalization was already highlighted and criticized earlier by Pavé (1989) and, more recently, by Dujarier (2017), who emphasizes the pretension of managers to design the organization. In this case, the qualifier proposed by Folcher (2015) " for the usage" applies in a relevant way, i.e. the devices are designed for a specific task (from a theoretical point of view) but are not necessarily correlated with the way the task is actually carried out.

– A computer-dominated mode, i.e. by what is technically feasible and not what is necessary for the users. The technical device will meet performance and technical criteria, but not criteria that are necessary for its use. In this perspective, importance is given to computer scientists and the technical possibilities they are able to put in place. In this configuration, misunderstandings may arise between the technical teams (who do not understand why their performant system is not used) and users who consider that the ITD does not take their needs into account.

– A usage mode, which is correlated with the needs of future users of the service. This mode feeds in particular on the development of user experience (UX) and design. Kankainen, in his research, defines UX as "the result of a motivated action in a certain context. Previous user experiences and expectations influence current experience, and current experience leads to other experiences and modified expectations" (Kankainen 2002, p. 31). Arhippainen and Tähti (2003) identify five categories of factors that need to be considered in a UX approach: social drivers, cultural drivers, user-specific factors, contextual factors and factors related to product characteristics. Indeed, with UX, the place given to the views of the end-users is preponderant and must make it possible to avoid a mismatch between them and the target system. We can illustrate UX with Figure 4.2, taken from the article by Barcenilla and Bastien (2009).

7 This trend is in line with the thinking of Taylorism and, more particularly, with the scientific organization of work (SOW) in which tasks are cut, timed and organized in such a way as to increase productivity.

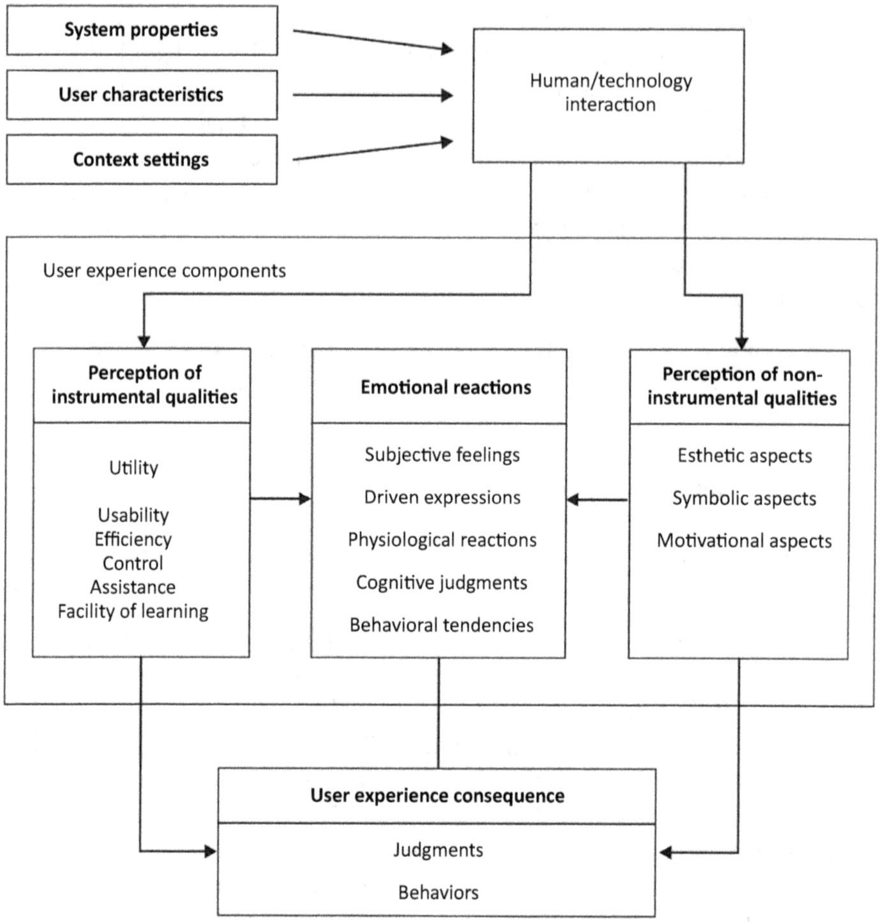

Figure 4.2. *Components of the user experience (according to Barcenilla and Bastien (2009, p. 325))*

COMMENT ON FIGURE 4.2.– *This schematic representation of user experience (UX) illustrates the different dimensions that are affected by UX. Thus, the factors related to user perception are taken into account, and also their aesthetic and semiotic universe and the way in which their emotions will be affected by the device. An ethical reflection*[8]*, relating to the influence that*

8 Constitution of several collectives of reflection on the subject such as Designers Éthiques (https://designersethiques.org/); organization of study days such as Ethics by Design

design can have on the behavior of the individuals, accompanies the work around the UX.

Thus, the sociotechnical systems identify and link "three modes of steering by organization, by information technology and by use [which] call, each time, for a distribution of competences and a distribution of specific authority" (Hochereau 2006, p. 291). From this triptych stems a situation that results (initially) in an inadequacy of the tools to the needs (piloting by the organization), towards a logic piloted by the technology and its possibilities which, with "designers struggling to validate the computer modeling they recommend, [and] users confronted with the incompleteness of the latter, lead to the alteration of the 'beautiful' scheduling initially programmed"[9] (Hochereau 2006, pp. 304–305). Finally, the last stage (user-driven) has the ambition to build both the organization and its instrumentation.

4.2.2. *Towards an instrumental genesis of the organization*

With the growing acceptance of the positioning of employees as users and potential designers of devices (in a logic of use) (Granjon and Denouël 2011), a consensus is emerging within organizations on the need to take into account employees as actors (Comtet 2007). Future users can engage in a process of co-creation of their system, which they think for their own use and considering their own functioning. The sociotechnical system then instruments and accompanies the changes in the organization through the reflective inclusion of transformations within it.

At a collective level, the users, by constructing their own device, implement the rules of the organization themselves and find themselves obeying the logic they have established, which echoes the teleological principle mentioned above. This point is to be taken into account at two levels:

– the user-designers are the first ambassadors of the use of the system to other employees. In this way, they help to promote the adoption of the tool, which then benefits from the legitimacy of usage;

(https://2018.ethicsbydesign.fr/); university work such as the work of the PROJEKT team of the University of Nîmes on design (http://projekt.unimes.fr/design/).

9 The author thus illustrates the situation of tension between a technical vision, on the one hand, and the users of the devices, on the other hand; this conjunction leads to non-use by users and a lack of understanding by technical teams even though the tool is technically very efficient.

– autonomy is accompanied, *de facto*, by self-submission, which translates into the commitment of the authors of the device to use it. By writing rules themselves into the system they develop, the designers, who are also the future users, implicitly accept to submit to these constraints.

By authorizing employees to develop their systems, the organization's hierarchy delegates by extension its control and organizational capacity (Appel and Heller 2010). We may then ask: does this way of operating lead to the optimization of the workforce or to the limitation of the political strength of employees? If the system can be qualified as strategic because of the stakes of power and the rules it imposes on its users, does its production by its future users change this characteristic? In other words, does the system contribute to greater freedom of action for employees? Or is it possible to better control the actions of employees by better framing their activity (Comtet 2006)? Does the organization thereby exercise greater control over employees than in the situation where they were forced to use a system imposed by management, but with the possibility of catachresis mentioned in Chapter 1? Indeed, the organization remains a space of power and domination. This is what Metzger (2011) points out when he states that managing to control change within the organization is a challenge in itself and that not all social actors have the necessary resources to intervene in decision-making.

From this perspective, by working on the genesis of devices, and more specifically on the instrumental genesis of a sociotechnical device (in Rabardel's sense), the instrumental genesis of organization is outlined, in which the device is inscribed by linking both instrumentalization and instrumentation. This organizational prism goes beyond the simple "user experience" to converge towards "organization experience"; it refers to *the collective capacity of the actors through their actions to contribute to the appropriation of the organization as an instrument through the deployment of their thought patterns.*

Thus, in the case of the construction of a specific device for a collective (a community of practice, for example), the future users contribute to the transformation of the organization by introducing a technological mediation of their collective activity (Lyytinen and Ngwenyama 1992). The instrumentation approach is part of a logic that is the opposite of that of constructing a device, since it is technically possible to do so. On the contrary, the device being built by the user-designer is centered on the actual uses and usage patterns implemented during its development. These usage patterns then move from an individual level to an organizational level by applying them to all the people working with the device in question.

Through this practical case, it is an instantiation of the "organization experience" that moves the actions of the members of the organization from individual action to collective action and where the evolution of the organization is accepted as the result of collective action. In this case, it is an opportunity to go further than Guyot (2000, p. 42) noted, since it is the collective that is at the origin of prescriptions, decisions and new organizational perspectives.

4.3. Organizational variations of collective intelligence

Collective intelligence, as presented in Chapter 3, extends to the "organization experience". If its emergence is based on sociotechnical devices, collective intelligence is then apprehended both in its technical dimension (through the mediation of sociotechnical instruments) and in its organizational dimension through its influence on the structure and function of the entity where it acts.

This work of reflection on the system is essential, since an information system is, above all, a symbolic system of representation. It can be mobilized in several ways: by the creation of entity–association diagrams or during transcriptions in the form of "life stories" (to describe the workings of a profession, for example) so that this system can be effectively implemented in the final system. The latter is then composed of the representations conceived and interpreted and links collective and/or individual actions by a technological base.

The role of the system is then to accompany the emerging collective intelligence; it is an effective response to the evolution of sectors that are becoming more decentralized and more "global"[10]. Through this emerging collective intelligence, organizations accentuate their reticulation and then tend to get closer to a networked entity, or even to the meta-organization presented in Chapter 3. The synthesis between sociotechnical digital

10 This decentralization can be observed both in the administrations, in connection with the decentralization laws (the main ones of which are accessible at this address: https://www.collectivites-locales.gouv.fr/principales-lois-decentralization), and in the industrial sector such as energy (with the production of so-called "clean" energy, which is produced locally before being reinjected into the national grid).

instrumentation and collective intelligence turns out to be a futuristic expression of the organization in the process of transformation towards a meta-organization.

4.3.1. *Collective intelligence and "glocal" logic*

The construction of the organization through collective action and collective intelligence makes it possible to concretely envisage an organizational change, as illustrated in Figure 4.3, in order to move from a positivist and hierarchical conception of the organization to a decentralized and deconcentrated organization, as mentioned above, based on a "glocal"[11] logic.

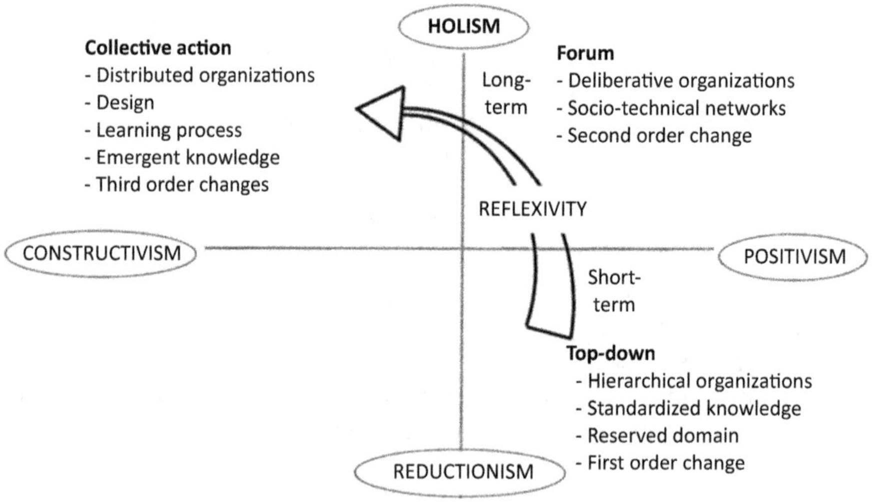

Figure 4.3. *Forms and modalities of action according to three points of view identified by Bawden (1997) and edited by Hubert (2005, p. 128)*

COMMENT ON FIGURE 4.3.– *This illustration represents three conceptions of the implementation of action within an organization. The question of the modality related to the action then reflects different postures. The lower right-hand quadrant reveals a production-oriented thinking, "techno-centric",*

11 The term "glocal" is a neologism that is a contraction of the terms global and local to mean the expression at a local level of a more global phenomenon. The local variant is then contextualized by its specificities that make its expression vary.

driven by standardization[12]*. The upper right-hand quadrant is said to be "eco-centric" by Hubert, i.e. in an ecosystem with a more reflexive vision (in particular on the fact that knowledge is not absolute). Finally, the insert marked by holism and constructivism "promotes social interactions and 'solidarities' (alliances as well as oppositions) ; it is no longer just a question of taking into account the existence of natural functions, but also the activities, practices, intentions and projects of human groups that draw their resources from these operations" (Hubert 2005, p. 147). A parallel can then be drawn between this thinking of "techno-centric"/"holistic" organization and the construction of devices centered on what it is technically possible to do/those centered on uses.*

As a result, the holistic organization (in Bawden's terminology, taken up by Hubert), based on digital tools, allows a rapid response to the permanent evolution of the ecosystem. It is then an effective solution to the instability of the environment in which organizations evolve. According to this holistic conception, the internal system of values adapts to change the current routines, rules or functioning inherent to it and can thus adjust to the current needs of the organization.

This functioning of the organization translates into a shift from self-centered directions to communities that network and collaborate, as illustrated in Figure 4.4.

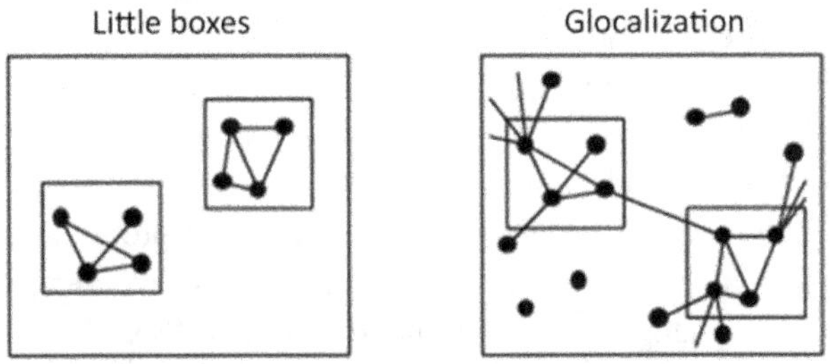

Figure 4.4. *Elevator and network communities (as cited in Wellman (2002, p. 13))*

COMMENT ON FIGURE 4.4.– *This representation illustrates two typologies of community functioning: a configuration called "little boxes", which is*

12 The organization will be structured either by the hierarchical links existing between employees or by the specialization of the employees. We then speak of a hierarchical or functional structure.

similar to silo structures[13]*, and a configuration called "glocalization", which transcends the silo partitioning to connect two collectives together. The representation that is made for this second conception recalls the concept of "structural hole" developed by Burt (1995), inspired by the work on the strength of weak ties theorized by Granovetter (1973).*

The *little boxes* configuration is very present in organizations (especially those considered in a functional or hierarchical way) where the different directions evolve in parallel, in silo. The author likens this situation to the relationships that the inhabitants of a village had and would be the configuration for which collaborative tools[14] have been developed (focusing in particular on the accomplishment of a task by a set of identified users).

In this case, the organization is closer to Bawden's top-down. Wellman's glocalization, on the other hand, relies on the collective action function represented by Hubert in Figure 4.2. From this perspective, it is the links between networks that tend towards a configuration of the organization based on collective action. According to Wellman, ICTs then facilitate this situation with the establishment of fluid systems facilitating access to individuals, materials or relationships. Individuals are able to simply change networks according to their needs (we find characteristics identified in Chapter 2 that take part in reticular organization).

Through collective action, the entire organization evolves: its functioning, culture and networks. Making different entities work together opens up opportunities: "the interactions between knowledge systems [...] open up new fields of action and relationships, which allow the emergence of new 'communities' of individual and collective actors linked by scientific and practical knowledge, techniques, norms, cultural preferences and natural realities [...] which then simultaneously make the knowledge, commitments and networks of exchange evolve" (Hubert 2005, p. 59). Within this holistic vision, the main issue is to provide the capacity to provide a response with a better consideration of the organizational dimension, i.e. to go beyond an individual vision to converge towards a more global vision (which echoes our vision of OX).

---

13 With, according to the author, people who belong to homogeneous groups (socially and cognitively).

14 Also known as Computer Support for Cooperative Work (CSCW).

These considerations resonate with the characteristics of collective intelligence that we have identified. Thus, a questioning transpires on the way collective intelligence can be set up, on the tools that participate in its constitution or even on its contribution to the organization, in the image of the question of knowledge.

It is then a breeding ground for the genesis of a collective intelligence marked by its local dimension. This change of perspective requires local knowledge and territorial intelligence (Moinet 2009; Pélissier 2009). The knowledge produced and used is then strongly marked by the culture of the region, and, mixed with the central culture, it can be considered as "glocal" knowledge. Moreover, the knowledge produced is contextualized by the environment in which the entity concerned evolves, which allows for decision-making that is more in tune with its ecosystem. What is important is the knowledge of the context in which the territory evolves and to be able to project itself over several years. We find the paradox where "the largest of networks paradoxically returns the smallest" (Cormerais 2015) in the continuation of Escarpit (1981) on the subject.

The observation made by Canet (2008) concerning social and political actors can be transcribed at the level of the organization where a "glocal" action then tends to make both the national and local visions converge. With communities, we are indeed in a situation where the legitimate hierarchical actor (such as a governing body) has the possibility of sharing decision-making power with other local actors who help him/her carry out his/her mission. For example, in the case of regional entities, the actors are as close as possible to the operational reality they understand and whose system of representation they share. The difficulty then lies in the organization's ability to adapt to local contexts while maintaining overall coherence.

Based on communities (of practice, virtual communities, and their synthesis, virtual communities of practice) and collective intelligence, the meta-organizational model offers a practical response to the needs of organizations and their functioning. These communities, in relation to other communities, benefit from a relative autonomy in the accomplishment of their missions and tend to use digital devices to exchange and organize. In this respect, the organizational model carried by collective intelligence contributes to the "organization experience" by promoting the expression of a collective construction linked to the technical dimension.

A complementary federating element between the members of the same entity is then knowledge, which can be understood as an immaterial common good in the logic of commons.

4.4. Intangible common goods

Collective action as well as the implementation of an organizational model based on meta-organization presupposes access to knowledge, information flows and internal data according to the needs of the stakeholders. The question of power games arises (Noyer 2016) in the access to these resources due to the assimilation of information to power (Tofler 1984; Chamoux 2013), all in a competitive and unstable context. Indeed, if it is possible to understand this intangible heritage as a common good that must be open to those who need it, the very high expectations of returns on investment and value creation through the "data" (Kitchin 2014) make it a strategic resource both internally and externally.

4.4.1. *Internal use of non-rival goods*

The term "common" can then be attributed to this heritage, which is shared by all members under the guise of knowing about it, knowing how to access it and being able to – indeed – access it (Cross *et al.* 2001).

This production of shared commons is reinforced by ICTs (De Moor 2011; Reed *et al.* 2014), since they improve the permeability between actors and the production of data or knowledge as commons that can be used in the principle of an internal "wisdom of the crowd" (Morell 2010; Last 2017). The application of the logic of "common goods" to the knowledge, data or information produced by the structure, however, calls into question the relevance of this approach. Indeed, Hardin (1968) illustrates the paradox with the metaphor of a system in which the individual tends to maximize his or her benefit by using the common good (in the case developed by Hardin, a field for grazing), and this in a context of a limited resource, a situation that leads, according to him, to overexploitation of the resource or even its drying up, with each person giving priority to his or her own interest.

Regarding the digital counterpart of commons, digital commons are considered as "information resources and knowledge that are collectively created and held or shared between or among a community and that tend to be non-excludable, i.e. (generally freely) accessible to third parties. As such, they are oriented to favor use and reuse, rather than exchange as a commodity. Moreover, the community of people that builds them can intervene in the governance of their interaction processes and shared resources" (Morell 2010, p. 5). However, unlike the material commons that Hardin mines, these immaterial commons have specific characteristics that Fulk and his co-authors present (Fulk *et al.* 2004):

– the individual action that makes up the collective action is based on the hypothesis that the increase in the number of contributions will lead to an increase in personal gain (like the way Wikipedia works) ;

– while the evaluation of the quantity and quality of tangible goods tends to be uniform among the members of a collective that uses them, for intangible goods, the evaluation of these same criteria can be very disparate depending on the users. Fulk *et al.* put forward three reasons for this: the fact that the number of contributors or the total volume is not known to everyone (the total number of contributors to a Wikipedia page is not always easy to obtain); that members do not all share the same vision of the maximum volume that can be reached; and finally that not all available data and information have the same value for people;

– the value of the intangible may decrease or increase by being shared, without dispossessing its original owner. Its financial or intellectual cost includes the work required to use the intangible asset; thus, simply copying and pasting or profoundly modifying the structure of the asset in order to use it changes its cost, as well as the specific skills required for its processing;

– an informational commons takes its value from the number of contributions and then from the quality and quantity that is perceived by the user;

– individual cost and value influence each other based on the person's interest in the informational property;

– by receiving an informational commons (which is potentially divisible), the individual adapts his/her behavior and perception of this good according to its perceived usefulness.

Immaterial production, considered as common, could, according to Hardin, avoid this "tragic" situation of material commons in the sense that data, as well as knowledge, are non-rival goods (Sammuelson 1954), i.e. it is possible to consume these goods without depriving another person of them. As described by Kitchin (2014), this non-rivalry is intrinsic to digital data; by consuming digital data, other potential consumers are always able to access it simultaneously.

4.4.2. *The use of the Web of Data for the provision of digital commons*

The challenge then arises: to enable all people with an information need to find the information that meets their need(s), while taking into account the proliferation of information systems – a situation that is not without complicating the search process (Passant 2009). Indeed, the information present in the organization must be usable by its members and stems from the information culture, which includes "the mastery of access to information [...], the culture of access to information [...], the information culture itself" (Juanals 2003, p. 24). It is possible to summarize these three concepts as "access and treatment of information, reasoned use of information, integration in a 'general culture' and in social uses" (Simonnot 2009, p. 29). The information culture in organizations is part of the framework of action, an important part of which is made up of informational activities (which are often invisible because of the integration of information and related activities into our daily lives).

The interoperability[15] of systems is therefore a major issue; in the case of non-interoperability of certain information systems, there are situations of double data entry, leading to loss of time and dissatisfaction on the part of operators. This loss of time is not in line with the logic of instantiality and the need for optimization for the organization (especially for companies); the

15 Interoperability is a notion composed of multiple facets, the most obvious of which is, *a priori*, according to Favier and Mustafa El Hadi (2013), the technical dimension. Indeed, interoperability allows the exchange of data between different systems (which is the foundation of the Internet, for example). In addition, there is also organizational interoperability (to facilitate work between different services or organizations) and semantic interoperability (to ensure the ability of systems to retain the meaning of the information exchanged).

organization then puts pressure on employees to optimize their time and to be part of the instantaneity allowed by ICTs.

The fragmentation of knowledge, spread across multiple IS, makes navigation and accessibility issues essential (Noyer 2016). A Web-based solution to this problem has been discussed for several years: the Semantic Web[16] (Berners-Lee *et al.* 2001), which tends to be assimilated more recently to the Linked Data Web[17] (Auer 2014) with an emphasis on data published in machine-accessible formats. In fact, based on the principle that web content is now available mainly in formats designed for humans (Web of Documents, also known as Web 1.0), the Web of Data allows the creation of a Web designed for machines. Linked data are described in a computer language with the structure of a triplet (Bachimont 2011): object (the resource to be described); properties (the properties of the resource); links (the connections between the resource and those attached to it). This treatment makes it possible to contextualize the data and thus avoid the situation mentioned by Worthy (2015), which is that of not being able to use the data for lack of context.

If the Web meets the first level of interoperability, called technical interoperability, which Favier *et al.* (2016) discusses, by promoting exchanges between technical platforms, the Web of Data tends towards the third level of interoperability that the authors identify: lexical and semantic interoperability. This level concerns "the ability to interpret the nature of the information exchanged" (Favier *et al.* 2016, p. 1), i.e. the description of information using, in particular, ontologies and semantic description

16 In 2001, the creator of the Web began to explore the possibilities related to the implementation of a so-called "Semantic" Web, thought as a technology that "will bring structure to the meaningful content of web pages, creating an environment in which software agents that move from page to page can easily perform sophisticated tasks for users" (Berners-Lee *et al.* 2001, p. 1). The stated goal is then to semantically enrich web pages to allow automatic processing of content that would be "understood" by programs.

17 The Linked Data Paradigm is a powerful way to transform the Web into the Web of Data and eventually into the Semantic Web. The approach has therefore evolved to maintain its momentum and to progressively achieve its goal by linking data (Dadzie and Rowe 2011). The *LOD-Cloud* illustrates this desire to link data together: https://lod-cloud.net/.

computer languages, such as the Simple Knowledge Organization System (SKOS)[18], to establish links similar in nature to those of thesauri.

Favier and Mustafa El Hadi already presented, in 2013, this issue of interoperability of systems (in particular knowledge organization systems, KOS) with a focus on semantic interoperability and, more particularly, on the "translatability of KOS" (Favier and Mustafa El Hadi 2013, p. 24) to reach a level where it is possible to be agnostic of the technical solution used. Indeed, by ensuring the semantic interoperability of several systems, the exchange of information will be transparent to the user. An illustration is the use of *owl:sameAs* to connect two resources together (Brickley and Miller 2010; Ristoski and Paulheim 2016) or the use of the "employee for" present in indexing and thesaurus languages[19]. The second level of interoperability, known as semantic, is based on "navigating through their diversity [of KOS]" (Favier and Mustafa El Hadi 2013, p. 24); indeed, there are currently many KOS and many approaches to describing data, and we must be able to take this heterogeneity into account.

The constitution of an information heritage accessible in real time and to which all members are connected is still utopian. However, the opportunities underlying this approach appear numerous, but they must be accompanied to avoid, in particular, the informational overflow, also known as "infobesity" (Mallowan 2012).

The functioning of the Web of Data then favors the interoperability of systems that can be part of this logic of widely accessible intangible commons. This technical brick then completes the other components mobilized through this work in order to finalize our framework of thought around the organization experience in the image of the Noosphere.

18 Information on the SKOS is available on the official website: https://www.w3.org/2004/02/skos/.

19 As in the "Répertoire d'autorité-matière encyclopédique et alphabétique unifié" (Rameau), the "Library of Congress Subject Headings" (LCSH) or "MOTBIS", to name but a few examples.

Conclusion

C.1. Organization and digital, between hybridization and calculability of reality

Organization is a research subject characterized by a diversity of approaches and currents of thought and by a multitude of concepts (communication of or within organizations, organizational theories, organizational knowledge, etc.). Caught in the net of the "information society" or the "knowledge society", the modern organization is undergoing many tensions and transformations generated by the increasing presence of digital technology.

In the continuity of business informatics, which has been deployed in organizations since 1980, the introduction of digital technology mediates the reality and is used in many personal and professional tasks. This trend continues the Enlightenment's desire to rationalize the world, and specifically the mathematization of the world of Leibnitz, through an ersatz that is instrumentation. The action of digital technology on reality allows its calculability (albeit still partial), a situation which is nowadays achieved by massive data processing technologies or by the predictive algorithms that feed off of them. The quantophrenia at work in our modern society is therefore a major trend where everything seems to be measurable and measured, if not understood.

Another illustration of this transformation of organizations is their transition to "2.0 organizations" (following the same logic as Web 2.0) by mobilizing the possibilities offered by digital technologies. Within these hybrid entities, both the immaterial and material dimensions of their activities coexist. Web 2.0 tools and technologies are similarly found in organizations, like the digital social networks found in intranets.

A type of reticular organization is unfolding, which is able to build on the opportunities of digital (such as telecommunication tools), and this up to its current most advanced form, the platform (which even affects the State, envisaged as "government as a service"). This organizational form, marked by its massive use of ICTs, is currently crystallizing contradictory visions with regard to the will underlying the creation of these platforms. Two basic trends can be distinguished:

– the first is marked by the thought of cybernetic governmentality, where the reticular organization must be optimized, in particular through algorithmic decision-making. It is accompanied by a platform capitalism, characterized by massive data collection, and the monitoring of the users of these platforms;

– the second is marked by the instrumentation of collective intelligence, so as to explore other forms of organization, such as the "holistic" organization proposed by Hubert or Wellman's "glocalization", which we carry through this work. The objective is then to promote "the socialization of problem solving rather than its treatment by separate bodies" (Lévy 1997, pp. 68–69).

C.2. The co-construction of sociotechnical devices...

Sociotechnical devices transcend technological or social determinisms, whose reciprocal influence – embodied by the current of sociotechnics – we accept by considering that in organizations, devices go beyond their technical dimension by being inscribed in a social context. Thus, they operate in a dynamic environment marked by values and social relations that contribute to their construction. However, in view of the choices made when they were designed, particularly if the system has been designed "for use" or "in use" (Folcher 2015), several contradictory logics are encountered in action situations.

The registration of these logics is manifested through the computer code written by the designers and developers of the future information system. The translation at work depends, therefore, largely on the actors mobilized for development. In this book, we have highlighted several repetitive situations in organizations that can be associated with the two logics mentioned above:

– logic "for use": we can group together the modes of conception thought in an organizational logic (symptomatic of management informatics) or by informatics (i.e. by what it is technically possible to do). The development by an external service provider or the implementation of a SaaS solution is mainly based on this logic. Indeed, the representation system will be the result of the service provider's understanding of the activity it formalizes in the source code. In the same way, if it is possible to set up a SaaS solution, the customization remains marginal or is carried out at a significant cost (and at the risk of incompatibility with future updates of the online solution);

– logic "in use": it is mainly the mode of design through use that is found in the stated objective of co-constructing the organization and its instrumentation. The development of in-house software solutions by actors not attached to the ITD is also an emanation of this approach. This "shadow IT" frequently responds to the needs expressed by operational teams that are not satisfied by solutions proposed at a higher level within the organization. In this case, the person in charge of development can share the end-users' usage patterns or at least access them more easily. The "in use" logic can create tensions between, on the one hand, experts with organizational legitimacy and, on the other hand, users with legitimacy through use. In addition to this, a network organization must ensure the coherence of its entire operation.

Concerning development by internal ITD teams, the outcome will depend heavily on the way the IT project is carried out (inclusion of future users or mobilization of experts). However, the introduction of agility and the cross-use of user experience favor the consideration of usage and future users to reduce the risks of rejection. Otherwise, users venture to divert the devices in question, which they are forced to use to respond to operational injunctions.

Because of this desire to co-construct, which can be explained by several factors (e.g. involving employees with the ambition of reducing the risk of IS rejection), we then act at several levels on the organization, such as temporality, which is influenced by the systems, or on the logic inherent in the usage patterns present in the systems. This evolution thus favors the sociotechnical mutation of the organization, which is transformed both technically and organizationally.

C.3. … to the co-transformation of the organization

We focus on the internal transformation of the organization with this concept of "organization experience", which is an approach to organizational design based on several pillars detailed in the book. Far from a cybernetic perspective, a vision embodied by the platform (or uberization[1]) or even the implementation of decentralized autonomous organization (DAO) like what is done on the blockchain[2], our vision is based on a logic of instrumental genesis of the organization.

Nourished by collective intelligence, the model that emerges in our proposal is that of a decentralized organization that allows for community-based problem solving and that brings together several intelligences within a single entity. Emphasis is then placed on the "glocal" dimension that collective intelligence can take on; the question of knowledge takes on its full importance here, even more so in the context of the "knowledge society" promoted by UNESCO. Indeed, knowledge permeates organizations that set up knowledge organization or knowledge management systems to ensure access, effective sharing or continuity (knowledge continuity). The increasing use of interoperable information systems favors this configuration, but many limitations still exist, such as the problems of system interoperability between several so-called "proprietary" environments. A current attempt to respond to these needs for sharing and access to knowledge is the

1 The control of drivers by Uber's algorithm illustrated by Rosenblat (2018) is here symptomatic of this logic of internal regulation of the system that aims to maintain its homeostasis (and, in this case, more precisely to increase its profits).

2 The example of *smart contracts*, which are "a more sophisticated device that allows the automatic execution of registered obligations as soon as the conditions for the execution of these commitments are met, taking into account all the conditions and limitations that were originally programmed into the contract. None of the parties may oppose performance" (Bar 2017, p. 75), is an illustration of the inherent governmentality of the blockchain (the *blockchain* in this case is Ethereum). These so-called "intelligent" contracts are only executed if specific criteria are validated. They will then deliver the programmed result.

construction of the Semantic Web (and its more recent declination as the Web of Data).

Digital technology has an impact on the organization and its functioning through the mediation it produces and its contribution to action. Its mediation is then a source of constraints and affluence that play on the three dimensions identified in Chapter 4:

– social: mediation through communication tools (such as digital social networks or instant messaging) or through interfaces to business processes;

– functional: indeed, as has been pointed out, the devices are binding by their nature (they transpose rules and standards into their code). They effectively transform the functioning of the organization[3]. Moreover, the possibilities offered, such as decentralization (which is facilitated), modify both the structure and the temporality of the organization (permanent beta logic, very regular updating or globalized work thanks to communication tools);

– epistemic: devices are means of action for the members of an organization and, by the same token, influence the representation of the organization that is produced. According to the present usage patterns, the resulting representations will be different, which influences the very nature of the organization.

We are, in fact, part of the transformation of organizations, with a particular link relating to digital, and emerging possibilities, but without limiting ourselves to this technical dimension. Figure C.1 summarizes the dimensions we are mobilizing in the context of OX.

Finally, we can highlight that the communicational and informational flows in the organization that structures them form part of the "symbolic cement", the importance of which Cordelier (2016) emphasizes in relation to maintaining the coherence of the organization. This is all the more important in an unstable context, characterized by technological ubiquity and decentralization favored by networked computing.

3 The deployment of new systems is almost systematically supplemented by a training plan and by a change support approach.

Figure C.1. *Summary diagram of the components of the organization experience mobilized throughout the work*

COMMENT ON FIGURE C.1.– *The components are complementary and independent even if they feed each other (e.g. collective intelligence feeds and promotes the emergence of a meta-organization) and converge towards an organizational transformation.*

Moving from a logic of user experience to that of organization experience is a way of associating both collective intelligence and digital instrumentation with a sociotechnical vocation in the organization. The considerations surrounding this "experience" are to be compared to the reading of experiential performance, where the ambition is to study the way in which employees become actors of organizational transformations and the potential conflicts with the existing order. These structural changes in the organization respond to environmental evolutions through the organizing dimension of symbolic action by means of the neologism "organizaction" by Morin (1990), which emphasizes, in particular, the capacity of self-organization of the organization through action.

Appendix

Framework for Analyzing the Potential of Collective Intelligence (Mačiulienė *et al.* 2015)

A.1. "Capacity" dimension

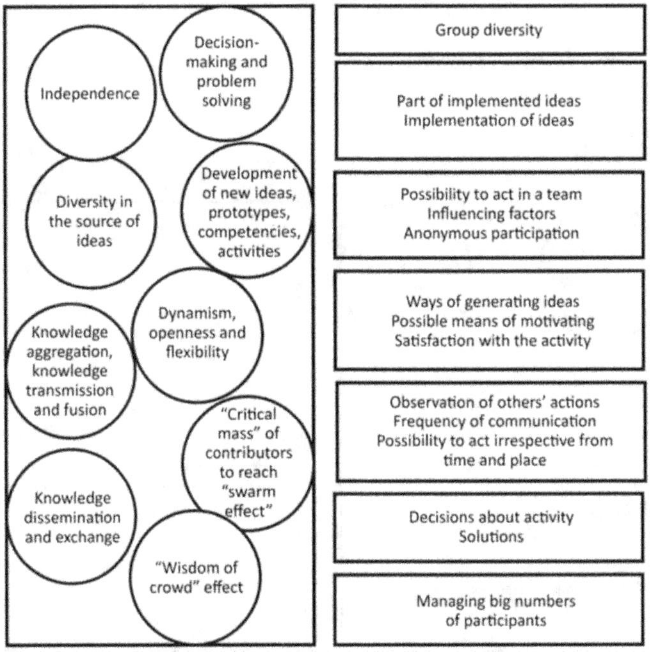

Figure A.1. *Details of the elements related to the "capacity" dimension of the framework of Mačiulienė* et al. *(2015)*

How useful is group diversity of an online community project?

Were the generated ideas implemented? In your opinion, what are the reasons of their implementation/non-implementation? Mention the stories of success and explain why they were "successful". How many generated ideas (what part) were implemented? What are the reasons of their implementation/non-implementation? Respondents were asked to mention the stories of success and explain why they were "successful".

How does such managerial feature of the project as the possibility to work in a team contribute to the project success? What has influence? Can you choose anonymous or public participation? Does it have any influence?

How are generated ideas selected? How are ideas filtered (is there a formal/informal procedure)? How are ideas classified? Who classifies them and in what way? How are different groups and strata involved, how are they motivated? What means of motivation could be used? What would help to involve more participants? Are you satisfied with the way activities/processes occur? Is everything developing as expected?

Is there a possibility to observe what others are doing? Online visibility option? How often do you communicate in a project? What influence does the activity of other group members have on participants? How do managerial features of the project (e.g. possibility to act at any time convenient for the participant, no time limits) contribute to project success?

Who decides project aims/activities/functions?
Are decisions made in a group of participants?

If the number of participants increased, would it be necessary to filter, classify, etc. the ideas?
Is the number of remaining participants relevant to you?

Figure A.2. *Issues associated with the capacity dimension*

A.2. "Emergence" dimension

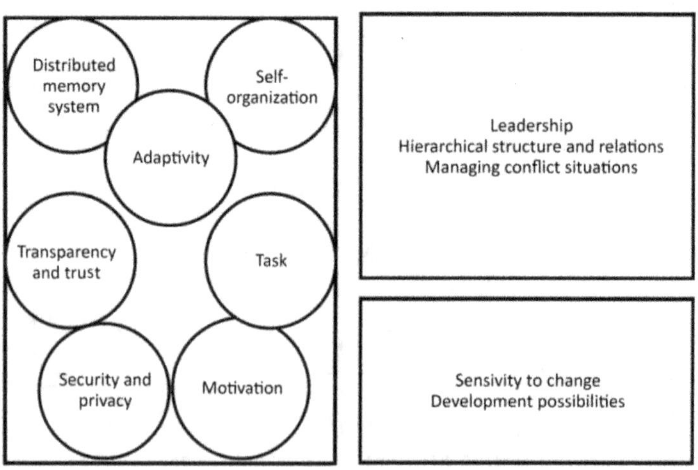

Figure A.3. *Details of the elements related to the "emergence" dimension of the framework of Mačiulienė* et al. *(2015)*

Are there any leaders and what are their functions?
How is group leaders' contribution assessed? How important is the number of the remaining active participants?
Hierarchical relations? Are there any? What hierarchical structure could be ideal?
What is the size of the group? What people participate (i.e. experience, age, education)? Have they participated in any other projects/communities? Geographical distribution of participants?
How could a conflict situation be resolved (e.g. blocking)? Who should do that?
What is done in the case of the confronting and harmful member of activity?

Are there any plans to change something? If not - why?
Do you have your internal criteria according to which you choose possible initiatives?
How different social problems are incorporated in communities' activities? Are certain subjects limited? Can all social challenges requiring a solution be proposed?

Figure A.4. *Issues associated with the emergence dimension*

A.3. "Social maturity" dimension

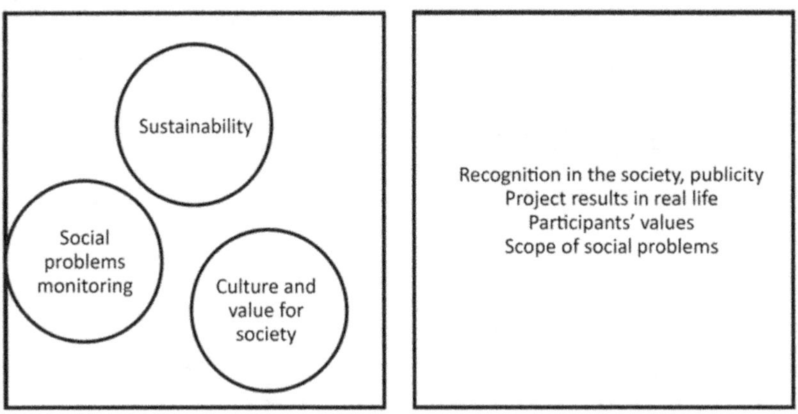

Figure A.5. *Details of the elements related to the "social maturity" dimension of the framework of Mačiulienė* et al. *(2015)*

How are different social problems incorporated in communities' activities?
Are you interested in social criteria published on the community website?
Are such criteria announced publicly?
Which social challenges do project initiators think they are solving?
Is it important to you what sources of financing projects receive? How much are project initiators dependent on those sources of financing?
What values do other project participants communicate?

Figure A.6. *Issues associated with the social maturity dimension*

References

Acquier, A., Daudigeos, T., and Pinkse, J. (2017). Promises and paradoxes of the sharing economy: An organizing framework. *Technological Forecasting and Social Change*, 125(C), 1–10.

Aggeri, F. (2014). Qu'est-ce qu'un dispositif stratégique ? Éléments théoriques, méthodologiques et empiriques. *AEGIS*, 1, 47–64.

Agostinelli, S. (2003). *Les nouveaux outils de communication des savoirs*. L'Harmattan, Paris.

Agostinelli, S. (2009). Comment penser la médiation inscrite dans les outils et leurs dispositifs. *Distances et savoirs*, 7, 355–376.

Ahrne, G. and Brunsson, N. (2005). Organizations and metaorganizations. *Scandinavian Journal of Management*, 21(4), 429–449.

Akrich, M. (1993). Les formes de médiation technique. *Réseaux*, 11(60), 87–98.

Akrich, M. (1998). Les utilisateurs, acteurs de l'innovation : l'utilisateur représenté. *Éducation permanente*, 134, 79–90.

Akrich, M., Callon, M., and Latour, B. (1988). À quoi tient le succès des innovations ? 2 : le choix des porteparole. *Gérer et comprendre*, 12, 14–29.

Alag, S. (2009). *Collective Intelligence in Action*. Manning Publications, New York.

Alauzen, M. (2019). L'état plateforme et l'identification numérique des usagers. *Réseaux*, 213, 211–239.

Appel, V. and Heller, T. (2010). Dispositif et recherche en communication des organisations. In *Les dispositifs d'information et de communication*, Appel, V., Boulanger, H., and Massou, L. (eds). De Boeck Université, Brussels.

Arazy, O., Morgan, W., and Patterson, R. (2006). Wisdom of the crowds: Decentralized knowledge construction in Wikipedia. In *16th Annual Workshop on Information Technologies & Systems (WITS) Paper*, Lu, Y. and Pant, G. (eds). WITS, Dublin.

Argyris, C. (2003). *Savoir pour agir*. Dunod, Paris.

Arhippainen, L. and Tähti, M. (2003). Empirical evaluation of user experience in two adaptive mobile application prototypes. *Proceedings of the 2nd International Conference on Mobile and Ubiquitous Multimedia (MUM)*. Linköping University, 27–34.

Aubert, N. (2003). *Le culte de l'urgence : la société malade du temps*. Flammarion, Paris.

Auer, S. (2014). Introduction to LOD2. In *Linked Open Data – Creating Knowledge Out of Interlinked Data*, Auer, S., Bryl, V., and Tramp, S. (eds). Springer International Publishing, Cham.

Autissier, D., Johnson, K.J., and Moutot, J.-M. (2016). L'innovation managériale : rupture ou évolution du management. *Question(s) de management*, 13(2), 25–33.

Bacache-Beauvallet, M., Bounie, D., and François, A. (2011). Existe-t-il une fracture numérique dans l'usage de l'administration en ligne ? *Revue économique*, 62(2), 215–235.

Bachimont, B. (2011). Enjeux et technologies : des données au sens. *Documentaliste-Sciences de l'Information*, 48(4), 24–41.

Ballay, J.-F. (2002). *Tous managers du savoir !* Éditions d'Organisation, Paris.

Balpe, J.-P., Saleh, I., and Lepage, D. (2003). *Hypertextes, hypermédias : créer du sens à l'ère numérique*. Hermes-Lavoisier, Paris.

Baltz, C. (1998). Une culture pour la société de l'information ? Position, définitions, enjeux. *Documentaliste-Sciences de l'Information*, 35(2), 75–82.

Barcenilla, J. and Bastien, J.M.C. (2009). L'acceptabilité des nouvelles technologies : quelles relations avec l'ergonomie, l'utilisabilité et l'expérience utilisateur ? *Le travail humain*, 72(4), 311–331.

Barreau, C. (2017). La régulation des smart contracts et les smart contracts des régulateurs. *Annales des mines – réalités industrielles*, 2017(3), 74–76.

Bastin, G. and Francony, J.-M. (2016). L'inscription, le masque et la donnée : datafication du web et conflits d'interprétation autour des données dans un laboratoire invisible des sciences sociales. *Revue d'anthropologie des connaissances*, 11(4), 505–530.

Batazzi-Alexis, C. (2002). Les technologies de l'information et de la communication (TIC) dans un processus d'apprentissage organisationnel : pour une coordination émergente entre le local et le global. *Communication et organisation*, 22, 1–12.

Battilana, J. and Lee, M. (2014). Advancing research on hybrid organizing – Insights from the study of social enterprises. *Academy of Management Annals*, 8(1), 397–441.

Baujard, C. (2011a). Conception et utilisation de la formation multimédia : quel dispositif juridique de protection des savoirs ? In *TIC et innovation organisationnelle*, Riccio, P.-M., Bonnet, D., and Perelman, J. (eds). Presses des Mines, Paris, 87–97.

Baujard, C. (2011b). Technologie organisationnelle et musées du futur : à la recherche d'un cadre managérial. In *TIC et innovation organisationnelle*, Riccio, P.-M., Bonnet, D., and Perelman, J. (eds). Presses des Mines, Paris.

Bauman, Z. (2006). *La vie liquide*. Actes Sud, Paris.

Bawden, R. (1997). Learning to persist: A systemic view of development. In *Systems for Sustainability: People, Organizations, and Environments*, Stowell, F.A., Ison, R.L., Armson, R., Holloway, J., Jackson, S., and McRobb, S. (eds). Springer, Boston.

Beckouche, P. (2017). La révolution numérique est-elle un tournant anthropologique ? *Le Débat*, 193(1), 153–166.

Bélisle, C. (2002). Médiatiser l'apprentissage aujourd'hui. In *Médiation, médiatisation et apprentissages*, Barbot, M.-J. and Lancien, T. (eds). ENS Éditions, Lyon.

Belton, L. and de Coninck, F. (2007). Des frontières et des liens. *Réseaux*, 2007/1(140), 67–100.

Benghozi, P.-J. (2000). Le développement des NTIC dans les entreprises françaises. Premiers constats. *Réseaux*, 18(104), 31–57.

Benghozi, P.-J., Pollet, P., Trahand, J., and Vargdanega-Lachaud, N. (2002). *Internet en réseau : au-delà de l'organisation hiérarchique et des technologies de demain*. L'Harmattan, Paris.

Bernard, F. (2000). Le lien communicationnel en organisation. *Sciences de la société*, 50–51, 25–45.

Bernard, F. (2004). Constructivisme et sciences de l'organisation. De l'alternative au pluralisme épistémologique "limité". *Communication et langages*, 139, 27–40.

Bernard, F. (2014). Imaginaire, participation, engagement et empowerment. *Communication et organisation*, 45, 87–98.

Bernard, F. (2015). La communication des organisations entre questions d'influence et questions d'autonomie. L'actualité des notions d'engagement, d'émergence et d'institution. *Communication et organisation*, 47, 85–95.

Bernard, F. (2016). Trente ans de recherches en communication des organisations : voies, paradoxes, imaginaires et questions vives. *Revue française des sciences de l'information et de la communication*, 9.

Berners-Lee, T., Hendler, J., and Lassila, O. (2001). The semantic web: A new form of web content that is meaningful to computers will unleash a revolution of new possibilities. *Scientific American*, 284(5), 35–43.

Berry, G. (2008). *Pourquoi et comment le monde devient numérique*. Collège de France, Paris.

Berthoud, G., Ischy, F., and Simioni, O. (2002). La société de l'information : la nouvelle frontière ? Report, Institut d'anthropologie et de sociologie, Lausanne.

Besson, P. (2016). Les ERP à l'épreuve de l'organisation. *Systèmes d'information & management*, 21(2), 17–47.

Besson, P. and Rowe, F. (2011). Perspectives sur le phénomène de la transformation organisationnelle. *Systèmes d'information & management*, 16(1), 3–34.

Bieber, M., Engelbart, D., Furuta, R., Hiltz, S.R., Noll, J., Preece, J., Stohr, E.A., Turoff, M., and Van de Walle, B. (2002). Toward virtual community knowledge evolution. *Journal of Management Information Systems*, 18(4), 11–35.

Bigham, J.P., Bernstein, M.S., and Adar, E. (2014). Human-computer interaction and collective intelligence. In *The Collective Intelligence Handbook*, Malone, T.W. and Bernstein, M.S. (eds). MIT Press, Boston.

Biron, M. and Hanuka, H. (2015). Comparing normative influences as determinants of knowledge continuity. *International Journal of Information Management*, 35(6), 655–661.

Blouin, A., Beaudoux, O., and Loiseau, S. (2009). Un tour d'horizon des approches pour la manipulation des données du Web. *Document Numérique*, 11(1), 63–83.

Blum, A., Hopcroft, J., and Kannan, R. (2016). *Foundations of Data Science*. Cornell CIS, New York.

Boboc, A. and Metzger, J.-L. (2009). Du privé vers le professionnel, une dynamique des apprentissages croisés autour des TIC. *Savoirs*, 20(2), 158–179.

Bonnafous-Boucher, M. and Dahl Rendtorff, J. (2014). *La théorie des parties prenantes*. La Découverte, Paris.

Bordier, R., Guérin, L., and Nussbaumer, J. (2012). Compte rendu du colloque sur l'ingénierie numérique organisé, le 25 novembre 2011, par l'Académie des Technologies, le Conseil Économique, Social et Environnemental et le Conseil Général de l'Industrie, de l'Énergie et des Technologies. *Annales des mines – réalités industrielles*, 2012(1), 117–125.

Borge-Holthoefer, J., Rivero, A., and Moreno, Y. (2012). Locating privileged spreaders on an online social network. *Physical Review E – Statistical, Nonlinear, and Soft Matter Physics*, 85(6), 1–7.

Botta-Genoulaz, V. (2007). Les ERP : un atout pour la cohérence, un risque pour la flexibilité ? In *La rationalisation dans les entreprises par les technologies coopératives*, de Terssac, G., Bazet, I., and Rapp, L. (eds). Octarès Editions, Toulouse.

Bouillon, J.-L. (2005). Autonomie professionnelle et rationnalisations cognitives : les paradoxes dissimulées des organisations post-disciplinaires. *Études de communications*, 28, 91–105.

Bouillon, J.-L. (2015). Technologies numériques d'information et de communication et rationalisations organisationnelles : les "compétences numériques" face à la modélisation. *Les enjeux de l'information et de la communication*, 16, 89–103.

Bouillon, J.-L., Bourdin, S., and Loneux, C. (2008). Approches communicationnelles des organisations : interroger l'organisation par la communication : éléments de contextualisation. *Sciences de la société*, 74, 3–9.

Bouveresse, J. (2002). Le mythe du progrès selon Wittgenstein et von Wright. *Mouvements*, 19(1), 126–141.

Brickley, D. and Miller, L. (2010). FOAF vocabulary specification. *Namespace Document*, 3.

Brooks, L., Atkinson, C., and Wainwright, D. (2008). Adapting structuration theory to understand the role of reflexivity: Problematization, clinical audit and information systems. *International Journal of Information Management*, 28(6), 453–460.

Brousseau, E. and Rallet, A. (1997). Le rôle des technologies de l'information et de la communication dans les changements organisationnels. In *Économie de la connaissance et organisation : entreprises, territoires, réseaux*, Guilhon, B., Huard, P., Orillard, M., and Zimmerman, J.B. (eds). L'Harmattan, Paris.

Burt, R.S. (1995). *Structural Holes: The Social Structure of Competition*. Harvard University Press, Cambridge.

Calude, C.S. and Longo, G. (2016). The deluge of spurious correlations in big data. *Foundations of Science*, 22(3), 595–612.

Canet, R. (2008). L'intelligence en essaim. Stratégie d'internationalisation des forums sociaux et régionalisation de la contestation mondiale. *Cultures & conflits*, 70, 33–56.

Carayol, V. and Bouldoires, A. (2011). *Discordance des temps : rythmes, temporalités, urgence à l'ére de la globalisation de la communication*. Maison des Sciences de l'Homme d'Aquitaine, Pessac.

Cardon, D. (1997). Présentation. *Réseaux*, 15(85), 5–10.

Cardon, D. and Casilli, A. (2015). *Qu'est-ce que le Digital Labor ?* INA, Paris.

Carmes, M. (2010). L'innovation organisationnelle sous les tensions performatives. Propositions pour l'analyse d'une co-construction conflictuelle des politiques et pratiques numériques. *Les cahiers du numérique*, 6(4), 15–37.

Carmes, M. (2018). *Digital Organizations Manufacturing: Scripts, Performativity and Semiopolitics*. ISTE Ltd, London and John Wiley & Sons, New York.

Carmes, M. and Noyer, J. (2014). L'irrésistible montée de l'algorithmique. Méthodes et concepts en SHS. *Les cahiers du numérique*, 10(4), 63–102.

Casilli, A. (2015). Digital labor : travail, technologies et conflictualités. In *Qu'est-ce que le Digital Labor ?* Cardon, D. and Casilli, A. (eds). INA, Paris.

Castagnos, J.-C. and Lesca, H. (2004). Capter les signaux faibles de la veille stratégique : retours d'expérience et recommandations. *E & G, Economia et Gestão*, 4(7), 15–34.

Casteignau, G. and Gonon, I. (2006). Pratique du travail collaboratif en communautés virtuelles d'apprentissage. *Hermès, La Revue*, 45, 109–115.

Castells, M. (1998). *La société en réseau*. Fayard, Paris.

Castro Goncalves, L., Chabault, D., and Tixier, J. (2011). Manager les tensions stratégiques entre le local et l'international : le cas des pôles de compétitivité. *Management & avenir*, 50(10), 172.

Chamoux, J.-P. (2013). Données publiques : un patrimoine commun ? *Les cahiers du numérique*, 9(1), 153–171.

de Chardin, P.T. (1995). *Le phénomène humain*. Le Seuil, Paris.

Chen, P.C.L. and Zhang, C.Y. (2014). Dataintensive applications, challenges, techniques and technologies: A survey on big data. *Information Sciences*, 275, 314–347.

Chevalier, A. and Kicka, M. (2006). Web designers and web users: Influence of the ergonomic quality of the web site on the information search. *International Journal of Human Computer Studies*, 64(10), 1031–1048.

Chou, S. and Hsu, C. (2018). An empirical investigation on knowledge use in virtual communities – A relationship development perspective. *International Journal of Information Management*, 38(1), 243–255.

Ciborra, C.U. (1996). The platform organization: Recombining strategies, structures, and surprises. *Organization Science*, 7(2), 103–118.

Cohendet, P., Simon, L., and Sarazin, B. (2017). *Les communautés d'innovation. De la liberté créatrice à l'innovation organisée*. EMS Éditions, Paris.

Collet-Thireau, K. and Thomas, J.-P. (2015). Big data et open data : quel impact pour les professionnels de l'information ? *I2D – Information, Données & Documents*, 53, 9–10.

Commandré, M. (2011). Synthèse des corrélations entre management des technologies organisationnelles et responsabilité sociale et environnementale. In *TIC et innovation organisationnelle*, Riccio, P.-M., Bonnet, D., and Perelman, J. (eds). Presses des Mines, Paris.

Compain, G., Eynaud, P., Morel, L., and Vercher-Chaptal, C. (2019). Alternative platforms and societal horizon: Characterisation and strategies for development. *SASE 31st Annual Meeting: Fathomless Futures: Algorithmic and Imagined.* SASE, New York.

Comtet, I. (2006). L'usage du groupware ou la construction d'un dispositif sociotechnique. *Revue française de gestion*, 168–169(9), 287–301.

Comtet, I. (2007). De l'usage des TIC en entreprise. Analyses croisées entre sciences de l'information et sciences de gestion. *Communication et organisation*, 31, 94–107.

Comtet, I. (2012). Les environnements collaboratifs de travail au service de l'intelligence collective économique ? *Communication et organisation*, 42, 62–72.

Cordelier, B. (2016). Retour sur le concept de transaction. *Revue française des sciences de l'information et de la communication*, 9.

Cormerais, F. (2015). L'hyperville : éléments pour un design territorial contributif et digital. *Traces numériques et territoires*, 161–165.

Coutant, A. (2015). Les approches sociotechniques dans la sociologie des usages en SIC. *Revue française des sciences de l'information et de la communication*, 6.

Cross, R., Parler, A., Prusak, L., and Borgatti, S. (2001). Knowing what we know: Supporting knowledge creation and sharing in social networks. *Organizational Dynamics*, 30(2), 100–120.

Crozier, M. and Friedberg, E. (1977). *L'acteur et le système*. Le Seuil, Paris.

Dadzie, A.S. and Rowe, M. (2011). Approaches to visualising linked data: A survey. *Semantic Web*, 2(2), 89–124.

Dameron, S. and Josserand, E. (2007). Le développement d'une communauté de pratique. Une analyse relationnelle. *Revue française de gestion*, 33(174), 131–148.

Davison, R.M. and Ou, C.X.J. (2017). Digital work in a digitally challenged organization. *Information and Management*, 54(1), 129–137.

Delcambre, P. (2017). Textes et échanges dans une organisation. *Communication et organisation*, 51, 123–140.

DeSanctis, G. and Poole, M.S. (1994). Capturing the complexity in advanced technology use: Adaptive structuration theory. *Organization Science*, 5(2), 121–147.

Domenget, J.-C. (2013). La fragilité des usages numériques. *Les cahiers du numérique*, 9(2), 47–75.

Domenget, J.-C., Miège, B., and Pélissier, N. (eds) (2017). *Temps et temporalités en information-communication : des concepts aux méthodes*. L'Harmattan, Paris.

Doueihi, M. (2008). *La grande conversion numérique*. Le Seuil, Paris.

Doueihi, M. (2011). *Pour un humanisme numérique*. Le Seuil, Paris.

Dubar, C. and Rolle, C. (2008). Les temporalités dans les sciences sociales : introduction. *Temporalités*, 8.

Dubé, L., Bourhis, A., and Jacob, R. (2005). The impact of structuring characteristics on the launching of virtual communities of practice. *Journal of Organizational Change Management*, 18(2), 145–166.

Dudezert, A. (2018). *La transformation digitale des entreprises*. La Découverte, Paris.

Dudezert, A., Fayart, P., and Oiry, E. (2014). Mythes-TIC et cultures, l'utopie de la gestion de la connaissance dans les organisations. *Revue internationale d'intelligence économique*, 6, 89–97.

Dudezert, A., Fayard, P., and Oiry, E. (2015). Astérix et la gestion des connaissances 2.0 : une exploration de l'appropriation des SGC 2.0 par le mythe du village gaulois. *Systèmes d'information et management*, 20(1), 31–59.

Dujarier, M.-A. (2017). *Le management désincarné : enquête sur les nouveaux cadres du travail*. La Découverte, Paris.

Dumas, A., Martin-Juchat, F., and Pierre, J. (2017). Les cadres et le numérique : cultures informationnelles et sentiment d'efficacité. *Communication et organisation*, 51, 55–65.

Durampart, M. (2009). Les TIC dans les organisations : des partenaires difficiles à contrôler. *Hermès, La Revue*, 54, 221–227.

Eberhard, C. (2013). *Oser le plurivers : pour une globalisation interculturelle et responsable*. Connaissances et savoirs, Paris.

Egnongo, Y. (2006). L'impact de la culture lors de l'importation d'une instrumentation de gestion des compétences dans une entreprise : cas de la GPEC dans un groupe bancaire africain. PhD Thesis, Université Grenoble Alpes, Grenoble.

El Amrani, R. (2008). De l'intégration du système d'Information à la vision transversale de l'organisation. *Systèmes d'information & management*, 13(4), 61–93.

Ély, F. (2010). L'analyse des pratiques info-communicationnelles dans les organisations : l'apport de l'école de Palo Alto. *Études de communication*, 35, 61–76.

Ermine, J.-L. (2003). *La gestion des connaissances*. Hermes-Lavoisier, Paris.

Escarpit, R. (1976). *Théorie générale de l'information et de la communication*. Hachette, Paris.

Escarpit, R. (1981). *Théorie de l'information et pratique politique*. Le Seuil, Paris.

Favier, L. and Mustafa El Hadi, W. (2013). L'interopérabilité des systèmes d'organisation des connaissances : une nouvelle conception de l'universalité du savoir ? In *Recherches ouvertes sur le numérique*, Papy, F. (ed.). Hermes-Lavoisier, Paris.

Favier, L., Mustafa El Hadi, W., and Vinck, D. (2016). Présentation : interopérabilité culturelle. *Communication*, 34(1).

Feeberg, A. (2014). Vers une théorie critique de l'Internet. *Tic & société*, 8(1–2), 31–56.

Ferrary, M. and Pesqueux, Y. (2004). *L'organisation en réseau, mythes et réalités*. PUF, Paris.

Fichez, E. and Varga, R. (2009). Construction d'un environnement numérique : la collaboration concepteurs/usagers. In *La communication éducative et les TIC : épistémologie et pratiques*, Sidir, M. (ed.). Hermes-Lavoisier, Paris.

Finchelstein, G. (2011). *La dictature de l'urgence*. Fayard, Paris.

Flichy, P. (2003). *L'innovation technique*. La Découverte, Paris.

Flichy, P. (2004). L'individualisme connecté entre la technique numérique et la société. *Réseaux*, 124, 17–51.

Flichy, P. (2008). Technique, usage et représentations. *Réseaux*, 2(148–149), 17–51.

Folcher, V. (2015). Conception pour et dans l'usage : la maîtrise d'usage en conduite de projet. *Revue des interactions humaines médiatisées*, 16(1), 39–60.

Fulk, J., Heino, R., Flanagin, A.J., Monge, P.R., and Bar, F. (2004). A test of the individual action model for organizational information commons. *Organization science*, 15(5), 569–585.

Gaglio, G. (2010). Pour une généralisation de l'innovation par l'aval dans les TIC. *Les cahiers du numérique*, 6(2), 19–37.

Gaglio, G. (2012a). Du lien entre l'analyse sociologique de l'innovation et la sociologie : une lecture simmelienne. *Cahiers de recherche sociologique*, 53, 49–72.

Gaglio, G. (2012b). *Sociologie de l'innovation*. PUF, Paris.

de Gail, M.-A. (2013). La ritualisation des interactions sur Facebook : cooptation et exposition de soi. *Les cahiers du numérique*, 9(3), 111–133.

Gallezot, G. and Le Deuff, O. (2009). Chercheurs 2.0 ? *Les cahiers du numérique*, 2(5), 15–31.

Galloway, A.R. (2012). *Les nouveaux réalistes : philosophie et postfordisme*. Léo Scheer, Paris.

Gangloff-Ziegler, C. (2009). Les freins au travail collaboratif. *Marché et organisations*, 10(3), 95–112.

Gardey, D. (2003). De la domination à l'action. Quel genre d'usage des technologies de l'information ? *Réseaux*, 120, 87–117.

Garud, R., Jain, S., and Tuertscher, P. (2008). Incomplete by design and designing for incompleteness. *Organization Studies*, 29(3), 351–371.

Gavillet, I. (2014). Michel Foucault et le dispositif : questions sur l'usage galvaudé d'un concept. In *Les dispositifs d'information et de communication*, Appel, V., Boulanger, H., and Massou, L. (eds). De Boeck Supérieur, Brussels, 17–38.

Geslin, P. (2002). Les formes sociales d'appropriations des objets techniques, ou le paradigme anthropotechnologique. *Ethnographiques.Org*, 1.

Gleick, J. (2015). *L'information : l'histoire, la théorie, le déluge*. Cassini, Paris.

Glenn, J.C. (2013). Collective intelligence systems and an application by the millennium project for the Egyptian academy of scientific research and technology. *Technological Forecasting and Social Change*, 97, 7–14.

Goëta, S. and Mabi, C. (2014). L'open data peut-il (encore) servir les citoyens ? *Mouvements*, 79(3), 81–91.

Granjon, F. (2009). Inégalités numériques et reconnaissance sociale. Des usages populaires de l'informatique connectée. *Les cahiers du numérique*, 5(1), 19–44.

Granjon, F. and Denouël, J. (2011). Penser les usages sociaux des technologies numériques d'information et de communication. In *Communiquer à l'ère numérique : regards croisés sur la sociologie des usages*, Denouël, J. and Granjon, F. (eds). Presses des Mines, Paris.

Granovetter, M. (1973). The strength of weak ties. *American Journal of Sociology*, 78, 1360–1380.

Greenan, N. and L'Horty, Y. (2000). Informatique, productivité et emploi : beaucoup d'espoirs, peu de certitudes. *Réseaux*, 18(100), 275–288.

Greenan, N. and Mairesse, J. (2006). Les changements organisationnels, l'informatisation des entreprises et le travail des salariés : un exercice de mesure à partir de données couplées entreprises/salariés. *Revue économique*, 57(6), 1137–1175.

Guez, J.-P. (2007). L'apport des SIC (Sciences de l'Information et de la Communication) pour une transformation des données et des acteurs du monde sportif. Un exemple : les entreprises équestres. *Staps*, 77(3), 95–106.

Guillemot, D. and Kocoglu, Y. (2010). Diffusion des outils dans les entreprises françaises. *Réseaux*, 4(162), 165–197.

Gulati, R., Puranam, P., and Tushman, M. (2012). Metaorganization design: Rethinking design in interorganizational and community contexts. *Strategic Management Journal*, 33(6), 571–586.

Guyot, B. (2000). Les dynamiques informationnelles. PhD Thesis, Université Grenoble III, Grenoble.

Guyot, B. (2006). *Dynamiques informationnelles dans les organisations*. Hermes-Lavoisier, Paris.

Guyot, B. (2009). Se mouvoir au sein du monde de l'information : comment les personnes parlent de leur activité d'information. *Études de communication*, 33, 101–118.

Guyot, B. (2013). Place et enjeux de l'information dans les organisations. In *L'information professionnelle*, Clavier, V. and Paganelli, C. (eds). Hermes-Lavoisier, Paris.

Hachour, H. (2011). Epistémologies socio-sémiotiques et communication organisante : la coproduction de sens comme moteur de l'organisation. *Communication et organisation*, 39, 195–209.

Hardin, G. (1968). The Tragedy of the Commons. *Science*, 162, 1243–1248.

Hassel, L. (2007). A continental philosophy perspective on knowledge management. *Information systems journal*, 17(2), 185–195.

Hassenzahl, M. (2008). User experience (UX): Towards an experiential perspective on product quality. In *Actes de la 20e Conférence Francophone sur l'Interaction Hommemachine*, Brangier, E. and Michel, G. (eds). Association for Computing Machinery, New York.

Hatchuel, A. (2005). Pour une épistémologie de l'action. L'expérience des sciences de gestion. In *Entre connaissance et organisation : l'activité collective*, Teulier, R. and Lorino, P. (eds). La Découverte, Paris.

Hénocque, B. (2002). *Appropriation des messageries électroniques dans les entreprises en réseau*. Presses Universitaires de Bordeaux, Pessac.

Henry, A. (2019a). Hybridation numérique d'une organisation énergétique. Analyse de son contexte de faisabilité sociotechnique. *Les cahiers du numérique*, 15(4), 113–136.

Henry, A. (2019b). L'émergence d'une intelligence collective dans une grande entreprise de l'énergie : étude de cas d'une communauté d'entraide œuvrant entre global et local. *Information, organisation, connaissances*, 2(1).

Heylighen, F. and Lenartowicz, M. (2017). The global brain as a model of the future information society: An introduction to the special issue. *Technological Forecasting and Social Change*, 114, 1–6.

Hobbes, T. (2000). *Léviathan ou matière, forme et puissance de l'état chrétien et civil*. Gallimard, Paris.

Hochereau, F. (2006). Concrétisation et normalisation. *Réseaux*, 1(135–136), 287–321.

Huang, J.S., Pan, S.L., and Liu, J. (2017). Boundary permeability and online-offline hybrid organization: A case study of Suning, China. *Information and Management*, 54(3), 304–316.

Hubert, B. (2005). L'interdisciplinarité sciences sociales/sciences de la nature dans les recherches sur problème. In *Entre connaissance et organisation : l'activité collective*, Teulier, R. and Lorino, P. (eds). La Découverte, Paris.

Illien, G., Hologne, O., Pouyllau, S., Alfonsi, G., Troeira, J.-P., Delahousse, J., Dalbin, S., von Rekowski, U., Aubry, C., and Huot, C. (2013). Enjeux professionnels. *Documentaliste-Sciences de l'Information*, 50(3), 26–41.

Jacquinot-Delaunay, G., and Monnoyer, L. (1999). Il était une fois. *Hermès, La Revue*, 25, 9–14.

Janssen, M. and Estevez, E. (2013). Lean government and platform-based governance: Doing more with less. *Government Information Quarterly*, 30, 1–8.

Jeanneret, Y. (2009). La relation entre médiation et usage dans les recherches en information-communication en France. *RECIIS*, 3(3), 1–9.

Jouët, J. (1993). Pratiques de communication et figures de la médiation. *Réseaux*, 11(60), 99–120.

Jouët, J. (2011). Des usages de la télématique aux Internet Studies. In *Communiquer à l'ère numérique : regards croisés sur la sociologie des usages*, Denouël, J. and Granjon, F. (eds). Presses des Mines, Paris, 45–90.

Juanals, B. (2003). *La culture de l'information. Du livre au numérique*. Hermes-Lavoisier, Paris.

Jung, J.J. (2017). Computational collective intelligence with big data: Challenges and opportunities. *Future Generation Computer Systems*, 66, 87–88.

Kankainen, A. (2002). Thinking model and tools for understanding user experience related to information appliance product concepts. PhD Thesis, Helsinki University of Technology, Helsinki.

de Kerckhove, D. (2000). *L'intelligence des réseaux*. Odile Jacob, Paris.

Kimble, C. and Hildreth, P. (2005). Dualities, distributed communities of practice and knowledge management. *Journal of Knowledge Management*, 9(4), 102–113.

Kitchin, R. (2014). *The Data Revolution: Big Data, Open Data, Data Infrastructures and their Consequences*. Sage, London.

Kiyindou, A. (2016). *Les sciences de l'information et de la communication par-delà les frontières*. L'Harmattan, Paris.

Koop, K. (2018). Monde plurivers et innovations sociales transformatrices. In *Les rencontres du pradel : transition récréative et écologie corporelle. Les cultures récréatives sontelles engagées sur le chemin de la transition ?* Le Pradel, Paris.

Lacombe, R., Bertin, P.-H., Vauglin, F., and Vieillefosse, A. (2011). Pour une politique ambitieuse des données publiques. Report, Ministère de l'Industrie, de l'Énergie et de l'Économie Numérique, Paris.

Laniau, J. (2009). Vers une nouvelle forme d'intelligence collective ? *Empan*, 76(4), 83–91.

Last, C. (2017). Global commons in the global brain. *Technological Forecasting and Social Change*, 114, 48–64.

Le Bon, G. (1905). *Psychologie des foules*. Félix Alcan, Paris.

Le Douarin, L. (2007a). C'est personnel ! *L'Homme et la société*, 163–164(1), 75–94.

Le Douarin, L. (2007b). Les chemins de l'articulation entre vie privée et vie professionnelle. *Réseaux*, 1(140), 101–132.

Lee, M.Y. and Edmondson, A.C. (2017). Self-managing organizations: Exploring the limits of less-hierarchical organizing. *Research in Organizational Behavior*, 37, 35–58.

Legrenzi, C. (2015). Informatique, numérique et système d'information : définitions, périmètres, enjeux économiques. *Vie & sciences de l'entreprise*, 200, 49–76.

Le Guel, F. (2004). Comment pourrait-on mesurer la double fracture numérique ? *Réseaux*, 5(127–128), 55–82.

Le Moënne, C. (2006). Quelques remarques sur la portée et les limites des modèles de communication organisationnelle. *Communication et organisation*, 30, 49–76.

Lenhardt, V. and Bernard, P. (2005). *L'intelligence collective en action*. Pearson France, Paris.

Lépine, V., Alemanno, S., and Le Moënne, C. (eds) (2017). *Communication & organisations : accélérations temporelles*. L'Harmattan, Paris.

Lévy, P. (1997). *L'intelligence collective : pour une anthropologie du cyberspace*. La Découverte, Paris.

Lévy, P. (2002). *Cyberdémocratie*. Odile Jacob, Paris.

Lévy, P. (2003). Le jeu de l'intelligence collective. *Sociétés*, 79, 105–122.

Lewandowski, A. and Bourguin, G. (2009). Quels supports informatiques pour un travail collaboratif durable ? *Marché et organisations*, 10(3), 155–170.

Liquète, V. (2012). Des pratiques d'information à la construction de connaissances en contexte : de l'analyse à la modélisation SEPICRI. PhD Thesis, Université de Rouen, Rouen.

Lloveria, V. (2014). Data design-moi un mouton. *Communication et organisation*, 46, 99–111.

Lombardo, É. and Angelini, C. (2013). Médiations mémorielles : le dispositif en tant que médiateur. *Communication et organisation*, 43, 193–204.

López-Huertas, M.J. (2016). The integration of culture in knowledge organization systems. *Advances in Knowledge Organization*, 15, 13–28.

Lorino, P. (2005). Théories des organisations, sens et action : le cheminement historique, du rationalisme à la genèse instrumentale des organisations. In *Entre connaissance et organisation : l'activité collective*, Lorino, P., and Teulier, R. (eds). La Découverte, Paris.

Lupia, A. and Sin, G. (2003). Which public goods are endangered? How evolving communication technologies affect the logic of collective action. *Public Choice*, 117(3–4), 315–331.

Lykourentzou, I., Vergados, D.J., Kapetanios, E., and Loumos, V. (2011). Collective intelligence systems: Classification and modeling. *Journal of Emerging Technologies in Web Intelligence*, 3(3), 217–226.

Lyytinen, K.J. and Ngwenyama, O.K. (1992). What does computer support for cooperative work mean? A structurational analysis of computer supported cooperative work. *Accounting, Management and Information Technologies*, 2(1), 19–37.

Maas, E. (2015). Culture managériale versus culture en action : les dimensions régulatrices des dynamiques culturelles dans les organisations. *Communication & management*, 12(2), 81–97.

Mačiulienė, M., Skaržauskienė, A., and Ribeiro-Navarrete, B. (2015). Emergence of collective intelligence in online communities. *Journal of Business Research*, 69(5), 1718–1724.

Mahé, S., Richard, B., Haik, P., Folino, A., and Musnik, N. (2010). Gestion des connaissances et systèmes d'organisation de connaissances : premier modèle et retours d'expérience industriels. *Document Numérique*, 13(2), 57–73.

Mair, J. and Reischauer, G. (2017). Capturing the dynamics of the sharing economy: Institutional research on the plural forms and practices of sharing economy organizations. *Technological Forecasting and Social Change*, July, 11–20.

Maleszka, M. and Nguyen, N.T. (2015). Integration computing and collective intelligence. *Expert Systems with Applications*, 42(1), 332–340.

Mallard, A. (2011). L'encadrement face au développement des interactions en réseau. Quelques réflexions sur le travail des managers dans les organisations fortement marquées par les TIC. In *TIC et innovation organisationnelle*, Riccio, P.-M., Bonnet, D., and Perelman, J. (eds). Presses des Mines, Paris.

Mallowan, M. (2012). Intelligence et transculture de l'information. *Communication et organisation*, 42, 27–48.

Manokha, I. (2018). Le scandale Cambridge Analytica contextualisé : le capital de plateforme, la surveillance et les données comme nouvelle "marchandise fictive". *Cultures & conflits*, 109, 39–59.

Marciniak, R. (2013). From organization design to meta organization design. In *Digital Enterprise Design and Management*, Benghozi, P.-J., Krob, D., and Rowe, F. (eds). Springer, Berlin.

Martino, L. (2011). Temps, essence des médias : l'actualité médiatique. In *Discordance des temps*, Carayol, V. and Bouldoires, A. (eds). Maison des Sciences de l'Homme d'Aquitaine, Pessac.

Martuccelli, D. (2016). L'innovation, le nouvel imaginaire du changement. *Quaderni*, 91, 33–45.

Mattelart, A. (2018). *Histoire de la société de l'information*. La Découverte, Paris.

McAfee, A. and Brynjolfsson, E. (2017). *Machine, Platform, Crowd: Harnessing our Digital Future*. W.W. Norton, New York.

McLuhan, M. and Powers, B. (1989). *The Global Village: Transformation in World Life and Media in the 21st Century*. Oxford University Press, New York.

McLure Wasko, M., and Faraj, S. (2000). "It is what one does": Why people participate and help others in electronic communities of practice. *Journal of Strategic Information Systems*, 9(2–3), 155–173.

Metzger, J.-L. (2011). TIC et travail. In *Communiquer à l'ère numérique : regards croisés sur la sociologie des usages*, Denouël, J. and Granjon, F. (eds). Presses des Mines, Paris.

Mevel, O. and Abgrall, P. (2009). Management de l'information dans l'organisation : une approche nouvelle de la veille informationnelle fondée sur le captage et le traitement des signaux faibles. *Revue internationale d'intelligence économique*, 1(1), 123–137.

Meyssonnier, F. and Pourtier, F. (2004). ERP, changement organisationnel et contrôle de gestion. In *Actes du 25e Congrès de l'Association Francophone de Comptabilité "Normes et Mondialisation"*, AFC (ed.). Association francophone de comptabilité, Orléans.

Miège, B. (2002). La société de l'information : toujours aussi inconcevable. *Revue européenne des sciences sociales*, XL-123, 41–54.

Miège, B. (2007). *La Société conquise par la communication : les Tic entre innovation technique et ancrage social*. Presses Universitaires de Grenoble, Grenoble.

Mohamad, A., Zainuddin, Y., Alam, N., and Kendall, G. (2017). Does decentralized decision making increase company performance through its information technology infrastructure investment? *International Journal of Accounting Information Systems*, 27(June), 1–15.

Mohamed, M.S. (2007). The triad of paradigms in globalization, ICT, and knowledge management interplay. *Vine*, 37(2), 100–122.

Moinet, N. (2009). Du "savoir pour agir" au "connaître est agir" : l'intelligence économique face au défi de la communication. *Les cahiers du numérique*, 5(4), 53–78.

de Moor, T. (2011). From common pastures to global commons: A historical perspective on interdisciplinary approaches to commons. *Natures Sciences Sociétés*, 19(2011), 422–431.

de Moor, A. and Weigand, H. (2007). Formalizing the evolution of virtual communities. *Information Systems*, 32(2), 223–247.

Morell, M.F. (2010). Governance of online creation communities: Provision of infrastructure for the building of digital commons. Thesis, European University Institute, Fiesole.

Morin, E. (1990). *Introduction à la pensée complexe*. Le Seuil, Paris.

Morin, E. (2015a). *L'aventure de la méthode*. Le Seuil, Paris.

Morin, E. (2015b). *Penser global: l'humain et son univers*. Robert Laffont, Paris.

Morozov, E. (2015). *Le mirage du numérique : pour une politique du big data*. Les Prairies ordinaires, Paris.

Morre, J.W. (2015). *Capitalism in the Web of Life: Ecology and the Accumulation of Capital*. Verso, London.

Mu, E., Kirsch, L.J., and Butler, B.S. (2015). The assimilation of enterprise information system: An interpretation systems perspective. *Information and Management*, 52(3), 359–370.

Muller, J.Z. (2018). *The Tyranny of Metrics*. Princeton University Press, Princeton.

Musso, P. (1998). *Télécommunications et philosophie des réseaux : la postérité paradoxale de Saint-Simon*. PUF, Paris.

Navarro, C. (2001). Partage de l'information en situation de coopération à distance et nouvelles technologies de la communication : bilan de recherches récentes. *Le travail humain*, 64(4), 297–319.

Ng, K.B. (2002). Toward a theoretical framework for understanding the relationship between situated action and planned action models of behavior in information retrieval contexts: Contributions from phenomenology. *Information Processing and Management*, 38(5), 613–626.

Ng, T. (2013). Information acquisition and institutions: An organizational perspective. *Information Economics and Policy*, 25(4), 301–311.

Nonaka, I. and Takeuchi, H. (1995). *The Knowledge-Creating Company: How Japanese Companies Create the Dynamics of Innovation*. Oxford University Press, Oxford.

Noyer, J.-M. (2016). *Transformation of Collective Intelligences*. ISTE Ltd, London and John Wiley & Sons, New York.

Ollion, E. and Boelaert, J. (2015). Au-delà des big data : les sciences sociales et la multiplication des données numériques. *Sociologie*, 6(3), 295–310.

Olson, G.M. and Olson, J.S. (2008). Computer-supported cooperative work. In *Handbook of Applied Cognition*, Durso, F.T. (ed.). Wiley, Hoboken.

O'Neil, C. (2018). *Algorithmes : la bombe à retardement*. Les Arènes, Paris.

O'Reilly, T. (2005). What is web 2.0: Design patterns and business models for the next generation of software [Online]. Available: http://oreilly.com.

O'Reilly, T. (2010). Government as a platform. In *Open Gouvernment*, Lathrop, D. and Ruma, L. (eds). O'Reilly Media, Sebastopol.

Orlikowski, W.J. (1992). The duality of technology: Rethinking the concept of technology in organizations. *Organization Science*, 3(3), 398–427.

Orlikowski, W.J. (1996a). Improvising organizational transformation over time: A situated change perspective. *Information System Research*, 7(1), 63–91.

Orlikowski, W.J. (1996b). Learning from notes: Organizational issues in groupware implementation. In *Computerization and Controversy*, Kling, R. (ed.). Academic Press, New York.

Pache, A.-C. and Santos, F. (2013). Inside the hybrid organization: Selective coupling as a response to competing institutional logics. *Academy of Management Journal*, 56(4), 972–1001.

Papy, F. (2009). *Technodocumentation des machines informationnelles aux bibliothèques numériques*. Hermes-Lavoisier, Paris.

Papy, F. (2016). *Digital Libraries: Interoperability and Uses*. ISTE Press, London, and Elsevier, Oxford.

Papy, F. and Sansonetti, D. (2014). *Les technologies du web au défi de l'entreprise*. Hermes-Lavoisier, Paris.

Paquienséguy, F. (2007). Comment réfléchir à la formation des usages liés aux technologies de l'information et de la communication numériques ? *Les enjeux de l'information et de la communication*, 1, 63–75.

Parker, G.G., Van Alstyne, M.W. and Choudary, S.P. (2016). *Platform Revolution: How Networked Markets are Transforming the Economy and How to Make Them Work for You*. W.W. Norton, New York.

Passant, A. (2009). Technologies du web sémantique pour l'Entreprise 2.0. PhD Thesis, Université Paris-Sorbonne, Paris.

Patil, D. and Mason, H. (2015). *Data Driven: Creating a Data Culture*. O'Reilly Media, Sebastopol.

Patrascu, M. (2010). Saisir les pratiques sociales du point de vue de leur organisation : revisiter le concept de médiation. *Les enjeux de l'information et de la communication*, 2, 58–70.

Pavé, F. (1989). *L'illusion informaticienne*. L'Harmattan, Paris.

Pélissier, M. (2009). Étude sur l'origine et les fondements de l'intelligence territoriale : l'intelligence territoriale comme une simple déclinaison de l'intelligence économique. *Revue internationale d'intelligence économique*, 1(2), 291–303.

Peng, R.D. and Matsui, E. (2015). *The Art of Data Science: A Guide for Anyone Who Works with Data*. Leanpub, Victoria.

Peraya, D. (1999). Médiation et médiatisation : le campus virtuel. *Hermès, La Revue*, 25, 153–167.

Perriault, J. (2002). *Éducation et nouvelles technologies : théorie et pratiques*. Nathan, Paris.

Perriault, J. (2008). *La logique de l'usage*. L'Harmattan, Paris.

Perriault, J. (2015). Retour sur la logique de l'usage. *Revue française des sciences de l'information et de la communication*, 6.

Pezziardi, P. and Verdier, H. (2017). *Des startups d'état à l'état plateforme*. Fondapol, Paris.

Pieters, W. (2013). On thinging things and serving services: Technological mediation and inseparable goods. *Ethics and Information Technology*, 15(3), 195–208.

Pitron, G. (2018). *La guerre des métaux rares : la face cachée de la transition énergétique et numérique*. Les Liens qui libèrent, Paris.

Plantin, J. and Valentin, J. (2013). Données ouvertes et cartographie libre. *Les cahiers du numérique*, 9, 85–107.

Polanyi, M. (1958). *Personal Knowledge. Towards a Post Critical Philosophy*. Routledge, London.

Polanyi, M. (1967). *The Tacit Dimension*. Anchor Books, New York.

Pontille, D. and Torny, D. (2013). La manufacture de l'évaluation scientifique : algorithmes, jeux de données et outils bibliométriques. *Réseaux*, 177(1), 23–61.

Prasarnphanich, P. and Gillenson, M.L. (2003). The hybrid clicks and bricks business model. *Communications of the ACM*, 46(12), 178–185.

Proulx, S. (2005). Penser les usages des technologies de l'information et de la communication aujourd'hui : enjeux, modèles, tendances. *Enjeux et usages des TIC : aspects sociaux et culturels*, 7–20.

Rabardel, P. (1995a). *Les hommes et les technologies : une approche cognitive des instruments contemporains*. Armand Colin, Paris.

Rabardel, P. (1995b). Qu'est qu'un instrument ? Appropriation, conceptualisation, mises en situation. *DIE*, 61–65.

Rabardel, P. (2005). Instrument, activité et développement du pouvoir d'agir. In *Entre connaissance et organisation : l'activité collective*, Lorino, P. and Teulier, R. (eds). La Découverte, Paris.

Rallet, A. (1986). La productivité ou l'impossible critère de gains réels apportés par la bureautique. *Réseaux*, 4(18), 45–63.

Rasse, P. (2015). Introduction : ce que les technologies font à la société. In *Sciences, techniques et société : rechercher sur les technologies digitales*, Masselot, C. and Rasse, P. (eds). L'Harmattan, Paris.

Reed, J.T., Jason, T., Smith, A., Parish, M., and Rickhoff, A. (2014). Using contemporary collective action to understand the use of computer-mediated communication in virtual citizen science. In *Online Collective Action: Dynamics of the Crowd in Social Media*, Agarwal, N., Lim, M., and Wigand, R.T. (eds). Springer, Vienna.

Reix, R. (2004). *Système d'information et management des organisations*. Vuilbert, Paris.

Rheingold, H. (1993). *The Virtual Community: Homesteading on the Electronic Frontier*. Harper Perennial, New York.

Rheingold, H. (2005). *Foules intelligentes : la révolution qui commence.* M2 Editions, Paris.

Ridings, C.M., Gefen, D., and Arinze, B. (2002). Some antecedents and effects of trust in virtual communities. *The Journal of Strategic Information Systems*, 11(3–4), 271–295.

Rifkin, J. (2000). *L'âge de l'accès : la révolution de la nouvelle économie*. La Découverte, Paris.

Rifkin, J. (2012). *La troisième révolution industrielle : comment le pouvoir latéral va transformer l'énergie, l'économie et le monde.* Les Liens qui libèrent, Paris.

Ristoski, P. and Paulheim, H. (2016). Semantic web in data mining and knowledge discovery: A comprehensive survey. *Web Semantics: Science, Services and Agents on the World Wide Web*, 36, 1–22.

Robert, P. (2009). *Une théorie sociétale des TIC. Penser les TIC entre approche critique et modélisation conceptuelle*. Hermes-Lavoisier, Paris.

Rochet, J.C. and Tirole, J. (2003). Platform competition in two sided markets. *Journal of the European Economic Association*, 1(4), 990–1029.

Rodota, S. (2011). Renforcement et développement de nouvelles applications informatiques de surveillance et de contrôle. *Terminal*, 108–109, 9–16.

Rosa, H. (2013). *Accélération : une critique sociale du temps*. La Découverte, Paris.

Rosenblat, A. (2018). *Uberland: How Algorithms Are Rewriting the Rules of Work.* University of California Press, Oakland.

de Rosnay, J. (1986). *Le cerveau planétaire.* Olivier Orban, Paris.

de Rosnay, J. (1995). *L'homme symbiotique : regards sur le troisième millénaire.* Le Seuil, Paris.

Rousseau, O., Gauthier, G., Saemmer, A., Tréhondart, N., Després-Lonnet, M., Mathis, R., Briatte, K., Bermès, E., and Pesquer, O. (2014). La documentation au service de la connaissance de l'œuvre. *Documentaliste-Sciences de l'Information*, 51, 56–67.

Roux, C. (2010). Les nouveaux modèles économiques des sociétés de l'Internet : les raisons de leur apparition. *Annales des mines – réalités industrielles*, 4, 42–47.

Ruijer, E., Grimmelikhuijsen, S., and Meijer, A. (2017). Open data for democracy: Developing a theoretical framework for open data use. *Government Information Quarterly*, 34(1), 45–52.

Saleh, I. and Hachour, H. (2012). Le numérique comme catalyseur épistémologique. *Revue française des sciences de l'information et de la communication*, 1.

Samuelson, P.A. (1954). The pure theory of public expenditure. *The Review of Economics and Statistics*, 36(4), 387–389.

Semprini, A. (2003). *La société de flux : formes du sens et identité dans les sociétés contemporaines*. L'Harmattan, Paris.

Serres, A. (2009a). Information culture. In *Information Science*, Papy, F. (ed.). ISTE Ltd, London and John Wiley& Sons, New York.

Serres, A. (2009b). Penser la culture informationnelle : des difficultés de l'exercice. *Les cahiers du numérique*, 5, 9–23.

Shan, C., Wang, H., Chen, W., and Song, M. (2015). *The Data Science Handbook*. Data Science Bookshelf, New York.

Shin, S.K. and Kook, W. (2014). Can knowledge be more accessible in a virtual network? Collective dynamics of knowledge transfer in a virtual knowledge organization network. *Decision Support Systems*, 59(1), 180–189.

Sieber, R.E. and Johnson, P.A. (2015). Civic open data at a crossroads: Dominant models and current challenges. *Government Information Quarterly*, 32(3), 308–315.

Silic, M. and Back, A. (2014). Shadow IT: A view from behind the curtain. *Computers and Security*, 45, 274–283.

Silic, M., Barlow, J.B., and Back, A. (2016). A new perspective on neutralization and deterrence: Predicting shadow IT usage. *Information and Management*, 54(8), 1023–1037.

Silva, F. and Ben Ali, A. (2010). Émergence du travail collaboratif : nouvelles formes d'organisation du travail. *Management & avenir*, 36(6), 340–365.

Simmel, G. (1988). Pont et porte. In *La tragédie de la culture et autres essais*, Simmel, G. (ed.). Rivages, Paris.

Simondon, G. (2014). *Sur la technique*. PUF, Paris.

Simonnot, B. (2009). Culture informationnelle, culture numérique : au-delà de l'utilitaire. *Les cahiers du numérique*, 5(3), 25–37.

Skaržauskienė, A., Pitrenaite-Zileniene, B., Leichteris, E., Paunksniene, Z., and Mačiulienė, M. (2014). Social technologies for developing collective intelligence in networked society. In *Collective Intelligence 2014*, Nickerson, J. and Malone, T. (eds). MIT Press, Cambridge.

Spender, J.C. and Marr, B. (2005). A knowledge-based perspective. In *Perspectives on Intellectual Capital*, Marr, B. (ed.). Sage, London.

Srnicek, N. (2018). *Capitalisme de plateforme : l'hégémonie de l'économie numérique*. Lux, Montreal.

Stacey, R.D. (1996). *Complexity and Creativity in Organizations*. Berrett-Koehler, San Francisco.

Stiegler, B. (2015). *La société automatique: l'avenir du travail*. Fayard, Paris.

Suchman, L.A. (1985). *Plans and Situated Actions: The Problem of Human-Machine Communication*. Xerox Palo Alto Research Center, Palo Alto.

Supiot, A. (2005). *Homo Juridicus : essai sur la fonction anthropologique du droit*. Le Seuil, Paris.

Surowiecki, J. (2004). *La sagesse des foules*. JC Lattès, Paris.

Svensson, G. (2001). "Glocalization" of business activities: A "glocal strategy" approach. *Management Decision*, 39(1), 6–18.

Szulanski, G. (1996). Exploring internal stickiness: Impediments to the transfer of best practice within the firm. *Strategic Management Journal*, 17, 27–43.

Tchuente, D. (2010). Pour une approche interdisciplinaire des TIC : le cas des réseaux socionumériques. *Document Numérique*, 14, 31–57.

Teigland, R. and Wasko, M.M. (2003). Integrating knowledge through information trading: Examining the relationship between boundary spanning communication and individual performance. *Decision Sciences*, 34(2), 261–287.

Tessier, N., Bourdon, I., and Kimble, C. (2014). Participer à une communauté de pratique virtuelle : retours d'expériences dans une multinationale de l'ingénierie. *Recherches en sciences de gestion*, 100(1), 121–140.

Teulier, R. (2005). Assister l'activité et l'organisation à travers des systèmes à base de connaissances. In *Entre connaissance et organisation : l'activité collective*, Lorino, P. and Teulier, R. (eds). La Découverte, Paris.

Thoemmes, J. (2008). Sociologie du travail et critique du temps industriel. *Temporalités*, 8, 1–11.

Thouin, M.F., Hoffman, J.J., and Ford, E.W. (2008). The effect of information technology investment on firm-level performance in the health care industry. *Health Care Management Review*, 33(1), 60–68.

Tisseron, S. (2001). *L'intimité surexposée*. Ramsay, Paris.

Toffler, A. (1984). *La troisième vague*. Denoël, Paris.

Toledo, C.M., Chiotti, O., and Galli, M.R. (2016). Process-aware approach for managing organisational knowledge. *Information Systems*, 62, 1–28.

Turner, F. (2012). *Aux sources de l'utopie numérique : de la contre-culture à la cyberculture, Stewart Brand, un homme d'influence*. C & F Éditions, Paris.

Vallon, R., da Silva Estácio, B.J., Prikladnicki, R., and Grechenig, T. (2017). Systematic literature review on agile practices in global software development. *Information and Software Technology*, 96, 161–180.

Varela, F. (1989). *Autonomie et connaissance : essai sur le vivant*. Le Seuil, Paris.

de Vaujany, F.-X. (2009). *Les grandes approches théoriques du système d'information*. Hermes-Lavoisier, Paris.

Vayre, J.-S. (2014). Manipuler les données. Documenter le marché. Les implications organisationnelles du mouvement big data. *Les cahiers du numérique*, 10(1), 95–125.

Vayre, J.-S. (2015). Les tableaux de bord sur données massives, pour un nouveau management de l'innovation ? *Innovations*, 47(2), 101.

Verdier, H. and Colin, N. (2012). *L'âge de la multitude : entreprendre et gouverner après la révolution numérique*. Armand Colin, Paris.

Verlaet, L. (2015). La deuxième révolution des systèmes d'information : vers le constructivisme numérique. *Hermès, La Revue*, 71(1), 249–254.

Walkowiak, E. (2006). Coordination, communication et modernisation des entreprises. *Hermès, La Revue*, 44, 57–64.

Wenger, E. (1998). *Communities of Practice*. Cambridge University Press, Cambridge.

Wenger, E., McDermott, R., and Snyder, W.M. (2002). *Cultivating Communities of Practice*. Harvard Business School Press, Brighton.

Wertsch, J. (1991). *Voices of the Mind: A Sociocultural Approach to Mediated Action*. Harvard University Press, Cambridge.

Wihlborg, E. and Söderholm, K. (2013). Mediators in action: Organizing sociotechnical system change. *Technology in Society*, 35(4), 267–275.

Winter, S., Berente, N., Howison, J., and Butler, B. (2014). Beyond the organizational "container": Conceptualizing 21st century sociotechnical work. *Information and Organization*, 24(4), 250–269.

Wolton, D. (2000). *Internet, et après ? Une théorie critique des nouveaux médias*. Flammarion, Paris.

Worthy, B. (2015). The impact of open data in the UK: Complex, unpredictable and political. *Public Administration*, 93(3), 788–805.

Yuhn, K.H. and Park, S.R. (2010). Information technology, organizational transformation and productivity growth: An examination of the Brynjolfsson-Hitt proposition. *Asian Economic Journal*, 24(1), 87–108.

Zacklad, M. (2016). Diversité des ontologies de la communication et de l'action collective. *Revue française des sciences de l'information et de la communication*, 9.

Zambon, S. and Monciardini, D. (2015). Intellectual capital and innovation. A guideline for future research. *Journal of Innovation Economics*, 17(2), 13–26.

Zhuge, H., Chen, J., Feng, Y., and Shi, X. (2002). A federation-agent-workflow simulation framework for virtual organisation development. *Information and Management*, 39(4), 325–336.

Zuboff, S. (1989). *In the Age of the Smart Machine*. Basic Books, New York.

Zuboff, S. (2019). *Age of Surveillance Capitalism: The Fight for a Human Future at the New Frontier of Power*. Public Affairs, New York.

Index

Other titles from

ISTE

in

Science, Society and New Technologies

2020

BARNOUIN Jacques
The World's Construction Mechanism: Trajectories, Imbalances and the Future of Societies
(Interdisciplinarity between Biological Sciences and Social Sciences Set – Volume 4)

ÇAĞLAR Nur, CURULLI Irene G., SIPAHIOĞLU Işıl Ruhi, MAVROMATIDIS Lazaros
Thresholds in Architectural Education (Engineering, Energy and Architecture Set – Volume 7)

ETCHEVERRIA Olivier
The Restaurant, A Geographical Approach: From Invention to Gourmet Tourist Destinations
(Tourism and Mobility Systems Set – Volume 3)

JEANNERET Yves
The Trace Factory
(Traces Set – Volume 3)

KATSAFADOS Petros, MAVROMATIDIS Elias, SPYROU Christos
Numerical Weather Prediction and Data Assimilation (Engineering, Energy and Architecture Set – Volume 6)

MARTI Caroline
Cultural Mediations of Brands: Unadvertization and Quest for Authority
(Communication Approaches to Commercial Mediation Set – Volume 1)

MAVROMATIDIS Lazaros E.
Climatic Heterotopias as Spaces of Inclusion: Sew Up the Urban Fabric
(Research in Architectural Education Set – Volume 1)

SCHMITT Daniel, THÉBAULT Marine, BURCZYKOWSKI Ludovic
Image Beyond the Screen: Projection Mapping

VIOLIER Philippe, with the collaboration of TAUNAY Benjamin
The Tourist Places of the World
(Tourism and Mobility Systems Set – Volume 2)

2019

BRIANÇON Muriel
The Meaning of Otherness in Education: Stakes, Forms, Process, Thoughts and Transfers
(Education Set – Volume 3)

DESCHAMPS Jacqueline
Mediation: A Concept for Information and Communication Sciences
(Concepts to Conceive 21st Century Society Set – Volume 1)

DOUSSET Laurent, PARK Sejin, GUILLE-ESCURET Georges
Kinship, Ecology and History: Renewal of Conjunctures
(Interdisciplinarity between Biological Sciences and Social Sciences Set – Volume 3)

DUPONT Olivier
Power
(Concepts to Conceive 21st Century Society Set – Volume 2)

FERRARATO Coline
Prospective Philosophy of Software: A Simondonian Study

GUAAYBESS Tourya
The Media in Arab Countries: From Development Theories to Cooperation Policies

HAGÈGE Hélène
Education for Responsibility
(Education Set – Volume 4)

LARDELLIER Pascal
The Ritual Institution of Society
(Traces Set – Volume 2)

LARROCHE Valérie
The Dispositif
(Concepts to Conceive 21st Century Society Set – Volume 3)

LATERRASSE Jean
Transport and Town Planning: The City in Search of Sustainable Development

LENOIR Virgil Cristian
Ethically Structured Processes
(Innovation and Responsibility Set – Volume 4)

LOPEZ Fanny, PELLEGRINO Margot, COUTARD Olivier
Local Energy Autonomy: Spaces, Scales, Politics
(Urban Engineering Set – Volume 1)

METZGER Jean-Paul
Discourse: A Concept for Information and Communication Sciences
(Concepts to Conceive 21st Century Society Set – Volume 4)

MICHA Irini, VAIOU Dina
Alternative Takes to the City
(Engineering, Energy and Architecture Set – Volume 5)

PÉLISSIER Chrysta
Learner Support in Online Learning Environments

PIETTE Albert
Theoretical Anthropology or How to Observe a Human Being
(Research, Innovative Theories and Methods in SSH Set – Volume 1)

PIRIOU Jérôme
The Tourist Region: A Co-Construction of Tourism Stakeholders
(Tourism and Mobility Systems Set – Volume 1)

PUMAIN Denise
Geographical Modeling: Cities and Territories
(Modeling Methodologies in Social Sciences Set – Volume 2)

WALDECK Roger
Methods and Interdisciplinarity
(Modeling Methodologies in Social Sciences Set – Volume 1)

2018

BARTHES Angela, CHAMPOLLION Pierre, ALPE Yves
Evolutions of the Complex Relationship Between Education and Territories
(Education Set – Volume 1)

BÉRANGER Jérôme
The Algorithmic Code of Ethics: Ethics at the Bedside of the Digital Revolution
(Technological Prospects and Social Applications Set – Volume 2)

DUGUÉ Bernard
Time, Emergences and Communications
(Engineering, Energy and Architecture Set – Volume 4)

GEORGANTOPOULOU Christina G., GEORGANTOPOULOS George A.
Fluid Mechanics in Channel, Pipe and Aerodynamic Design Geometries 1
(Engineering, Energy and Architecture Set – Volume 2)

GEORGANTOPOULOU Christina G., GEORGANTOPOULOS George A.
Fluid Mechanics in Channel, Pipe and Aerodynamic Design Geometries 2
(Engineering, Energy and Architecture Set – Volume 3)

GUILLE-ESCURET Georges
Social Structures and Natural Systems: Is a Scientific Assemblage Workable?
(Social Interdisciplinarity Set – Volume 2)

LARINI Michel, BARTHES Angela
Quantitative and Statistical Data in Education: From Data Collection to Data Processing
(Education Set – Volume 2)

LELEU-MERVIEL Sylvie
Informational Tracking
(Traces Set – Volume 1)

SALGUES Bruno
Society 5.0: Industry of the Future, Technologies, Methods and Tools
(Technological Prospects and Social Applications Set – Volume 1)

TRESTINI Marc
Modeling of Next Generation Digital Learning Environments: Complex Systems Theory

2017

ANICHINI Giulia, CARRARO Flavia, GESLIN Philippe,
GUILLE-ESCURET Georges
Technicity vs Scientificity – Complementarities and Rivalries
(Interdisciplinarity between Biological Sciences and Social Sciences Set – Volume 2)

DUGUÉ Bernard
Information and the World Stage – From Philosophy to Science, the World of Forms and Communications
(Engineering, Energy and Architecture Set – Volume 1)

GESLIN Philippe
Inside Anthropotechnology – User and Culture Centered Experience
(Social Interdisciplinarity Set – Volume 1)

GORIA Stéphane
Methods and Tools for Creative Competitive Intelligence

KEMBELLEC Gérald, BROUDOUS EVELYNE
Reading and Writing Knowledge in Scientific Communities: Digital Humanities and Knowledge Construction

MAESSCHALCK Marc
Reflexive Governance for Research and Innovative Knowledge
(Responsible Research and Innovation Set - Volume 6)

PARK Sejin, GUILLE-ESCURET Georges
Sociobiology vs Socioecology: Consequences of an Unraveling Debate
(Interdisciplinarity between Biological Sciences and Social Sciences Set – Volume 1)

PELLÉ Sophie
Business, Innovation and Responsibility
(Responsible Research and Innovation Set – Volume 7)

2016

BRONNER Gérald
Belief and Misbelief Asymmetry on the Internet

EL FALLAH SEGHROUCHNI Amal, ISHIKAWA Fuyuki, HÉRAULT Laurent, TOKUDA Hideyuki
Enablers for Smart Cities

GIANNI Robert
Responsibility and Freedom
(Responsible Research and Innovation Set – Volume 2)

GRUNWALD Armin
The Hermeneutic Side of Responsible Research and Innovation
(Responsible Research and Innovation Set – Volume 5)

LAGRAÑA Fernando
E-mail and Behavioral Changes: Uses and Misuses of Electronic Communications

LENOIR Virgil Cristian
Ethical Efficiency: Responsibility and Contingency
(Responsible Research and Innovation Set – Volume 1)

MAESSCHALCK Marc
Reflexive Governance for Research and Innovative Knowledge
(Responsible Research and Innovation Set – Volume 6)

PELLÉ Sophie, REBER Bernard
From Ethical Review to Responsible Research and Innovation
(Responsible Research and Innovation Set – Volume 3)

REBER Bernard
Precautionary Principle, Pluralism and Deliberation: Sciences and Ethics
(Responsible Research and Innovation Set – Volume 4)

VENTRE Daniel
Information Warfare – 2nd edition

www.ingramcontent.com/pod-product-compliance
Lightning Source LLC
LaVergne TN
LVHW011804200926
848484LV00003B/782